Human Genetics

Human Genetics

A. M. Winchester
University of
Northern Colorado

Charles E. Merrill Publishing Company
A Bell & Howell Company
Columbus, Ohio

International Standard Book Number: 0–675–09206–X

Library of Congress Catalog Card Number: 79–153769

3   4   5   6   7   8—76   75   74   73   72
Printed in the United States of America

PREFACE

Everyone is vitally concerned with human genetics. So much of what we are, what we do, and what we can transmit to future generations is related to our genes. Many of our social, medical, and personal problems have their roots in genetics, and a better understanding of genes and how they operate can aid in solving these problems.

Knowledge in this field has mushroomed in recent years. Genetics began as a rather vague concept about a mysterious force which caused children to be like their parents. Today there is a detailed understanding of the nature of hereditary material, how it operates, and how to thwart some of its harmful effects. People who only a few years ago would have been doomed to a life in a mental institution because of inherited defects are now leading normal lives. And those who inherited genes which would have brought about blindness, crippled muscles, and early death are now free of these harmful effects. This is what is happening today, and the future holds even greater promise.

It is the purpose of this book to present some of these findings at a level which can be understood by the average reader with intellectual curiosity. The author wrote this material primarily as a supplement to a first course in biology. No prerequisites are necessary; language, terms, and statistics have been simplified or eliminated with the goal of providing a readable and thought provoking book. The experienced human geneticist who reads this book will realize that many small exceptions and detailed technicalities have been omitted. Space prevents the inclusion of all these in a book of this size and at this level. Those who wish to pursue the subject further can do so through more advanced study.

The book is liberally illustrated with drawings and photographs. Such illustrations can often clarify points which are difficult to describe in words. Sincere appreciation is due the editors, artists, and production personnel of the Charles E. Merrill Publishing Company for their efforts to make this a book which is attactive as well as informative.

A. M. Winchester
May, 1971

CONTENTS

Chapter 1

HOW HEREDITY AFFECTS
YOUR LIFE

When you look into a mirror you see features which are distinctive. No other person on earth has exactly the same combination of facial characteristics as you possess. Your friends can recognize you at a glance. Only if you have an identical twin is there likely to be any confusion. What determined the many facial features and expressions which make you one of a kind in a world of over three billion people? It is your genes, those tiny units of heredity which are found by the thousands in the cells of your body. No one else has the particular combination of genes which you have (identical twins excepted). These genes determined how your body was constructed from the food which you absorbed from your mother before birth and the food you ate after birth. They have also determined many less tangible characteristics, such as your aptitudes in various types of endeavors, your resistance to certain diseases, the chemical structure of your blood, and how you will react to various environmental stimuli. They are now functioning in your body, producing many of the vital materials needed for living, and will continue functioning throughout your life.

Nothing is as important to you as your genes. They give you all of your potentialities; you cannot go beyond the limitations imposed by your genes. This is not to belittle the impor-

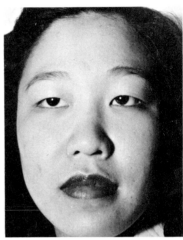

FIGURE 1-1

Facial features show great variation according to genetic background. These two young ladies have inherited features characteristic of different races, but each also has a gene combination from within her race which makes her unique. No other person on earth looks just like either one of these women.

tance of environment. Genes build from the materials supplied to them. They cannot do their best job unless they are supplied with the materials they need. Amino acids, vitamins, and certain inorganic minerals are essential if genes are to function at their best. Use also plays an important part. You may inherit a potential for well-formed legs, but if your legs are not used to any great extent, they will be underdeveloped and poorly formed. Likewise, you may inherit a brain with a great potential, but if this brain is not used, it will fall far below its potential.

GENES, THE MASTER PLANS

Did you ever wonder why a female dog never gives birth to kittens instead of puppies? Suppose you have a female dog and a female cat, both of which are soon to give birth to offspring.

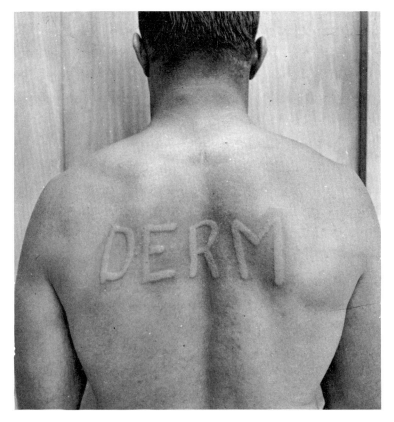

FIGURE 1-2

Many unusual traits appear as a result of the action of the hereditary genes. One of these is dermatography, the ability to write on the skin. The skin responds to moderate pressure by swelling due to accumulation of tissue fluid at the point of pressure. This photograph shows the result of tracing letters on the skin with the blunt end of a pencil.

You feed both of them exactly the same food, yet puppies are produced in one case and kittens in the other. How can this marvel come about? Genes make the difference; both a puppy and a kitten start life as a single cell, a fertilized egg. If you examined the fertilized eggs of a cat and a dog, even with a high powered microscope, you could not tell which would form the puppy nor which would form the kitten. They would look alike to you, yet they differ in one important respect. One contains genes which can make a dog, and the other contains genes which can make a cat. When food is eaten by the female dog, it is digested—that is it is broken down into soluble particles which can be absorbed and carried by the blood. When some of this digested food reaches an embryo, it is rearranged in such a way as to form tissues which are characteristic of a dog. The genes continue their work after the puppy is born, and finally a full-grown dog will be formed. This will not only be a dog, but a dog which is of the same general breed as its parents and which has many individual characteristics which were found in its parents.

If we could find some way to transplant the fertilized eggs from a cat into the body of a female dog and get implantation, the dog would undoubtedly give birth to kittens. Such transplants are done in cattle today. An egg can be taken from a fine breed of a cow and fertilized in a test tube with sperms of a bull from the same breed. The young embryo can then be placed in the body of some scrub cow with practically no commercial value. Yet the calf which is born will have the same fine qualities of the pure breed, just as if it had developed in the body of the cow from which the egg was taken. It is not the place of development, but the genes contained in the cells which determine what type of animal will develop.

The same principles hold true for the development of human life. The family cat and the family dog may receive the scraps from the food on the table of a prospective human mother, yet the food eaten by the woman is eventually used by the human embryo to form human skin, bone, muscle, and other tissue so that a human baby is born and not a kitten or a puppy. It may be hard to imagine that the huge fullback that you see crashing through the line at a football game was once a tiny baby. Yet he was, and even earlier he was only a small blob of protoplasm absorbing food from his mother's body. His genes made possible the great mass of coordinated bone and muscle that you see making a big gain on the football field.

Genes are like plans used by builders to construct a house. The builders take an assortment of lumber, bricks, mortar, glass, nails, and other materials. They follow a set of plans and put all of these things together to form a certain kind of house. The same builders, following a different set of plans, might take the same materials and produce a house which was entirely different. The genes furnish the plans for the construction of living beings. The genes in the reproductive cells of a cat have a different set of plans from those found in human reproductive cells.

FIGURE 1-3

What a difference genes can make. Both of these babies were constructed from the same raw materials; it was only the difference in the genes of the two fertilized eggs which caused these materials to be put together in the shape of a baby kitten in one case and a human baby in the other.

SOME STRANGE BELIEFS ABOUT HEREDITY

People have always been interested in heredity. When children show traits of their parents, it is quite evident that there must be some link which could not be accounted for on the basis of environment alone. When people have no satisfactory explanation for an observation, they are very likely to make guesses and sometimes come to believe their guesses even though they have no proof. One of the older assumptions about heredity, which many people came to believe, holds that traits we have acquired as a result of adaptation to the environment are passed on to our children. A woman with a fair skin may develop a suntan if she is exposed to the sun over a period of time, but this does not mean that her children will have dark skins as a result of her exposure. A man who uses his arms a great deal will develop powerful muscles in his arms, but this does not mean that his children will have larger arm muscles as a consequence of their father's activities. Many old beliefs about heredity included the inheritance of acquired characteristics. There was even a theory to explain such inheritance. This theory held that tiny particles moved down from all parts of the human body to the reproductive cells. When these reproductive cells were used in the formation of a new life, the tiny particles were supposed to make the parts of the body of the embryo like the parts of the body from which they came. In other words, particles from the strong man's arms moved to his reproductive cells and made strong arms in his children. It was even suggested that we could develop a race of one-eyed people if one eye of each baby was destroyed as soon as the baby was born.

Careful studies of heredity have shown that such a method of transmission of characteristics does not exist. The genes are responsible for heredity, and genes are not changed by the characteristics a person develops as a result of his individual adaptations.

Another old belief held that blood was the agent of heredity and that children were actually produced by a mingling of the blood of the two parents. Royal families believed that they had some special kind of blood, different from common blood, and therefore royal persons were allowed to marry only other royal persons so there would be no dilution of the royal blood. Even today some people do not like to take a blood transfusion for fear that they might become like the person who donated the blood. We know today that blood is not the material of heredity, and no person develops any traits as a result of blood transfusion. We still use terms from the past, however, when we refer to heredity. We speak of a "blood relative" when we mean someone related to us directly rather than by marriage. When someone gets into trouble, we speak sometimes of "bad blood" when we wish to imply that heredity may have played a part in the wrongdoing.

Can a woman mark a child or cause its heredity to be changed while she is carrying the child before birth? This was a very prevalent belief in the past, and there are still many people today who feel that there is some truth in it. The idea is that when a woman receives some great shock or other strong mental impression, the embryonic child in her body can somehow be changed because of this impression. For example, one woman saw a serious automobile accident while she was carrying a child. Later, the child was born without an arm, and she believed that the shock she received when she saw the accident caused the child to be so marked. There are many such stories; a woman is frightened by a frog that jumps at her unexpectedly, and the child is born with a birthmark which looks like a frog, or she may crave strawberries, and the child has a stawberry-shaped birthmark.

Modern knowledge shows that such events do not mark a child. Certainly every woman contacts many things while she is carrying the child, and if the child should chance to show some sort of abnormality, it is easy to find some event which seems to have some connection with the abnormality. There is no way in which such events can change the child; however, and the connection is coincidental. This does not mean that the mother has no influence on the child she is carrying. The food she eats goes to nourish the child, and if she has an inadequate diet, the embryo will not be nourished properly and will not be as strong and healthy when born as if the mother had been well nourished. Or, if the woman is exposed to certain types of radiation which penetrate her body, or takes certain drugs, or has certain diseases, the embryo may be damaged or abnormal when born. These effects, however, are quite different from the old idea that a woman's mental impressions could create some mark on the embryo.

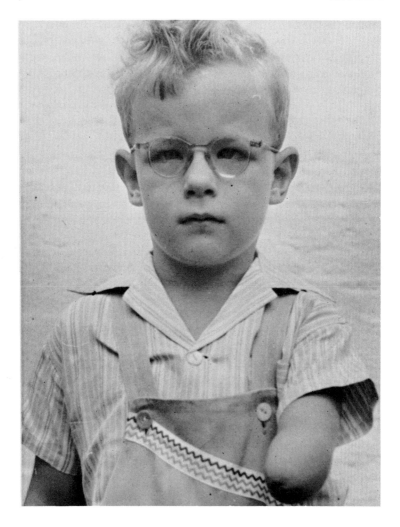

FIGURE 1–4

A maternal impression? The mother felt that the defective arm of her child resulted from a strong impression she received during her pregnancy when she saw people injured in an automobile accident. Actually, the inheritance of a rare gene caused the defect. The old belief in such maternal impressions has been disproved.

Chapter 2

THE BRIDGE OF LIFE

Every human being started life as a tiny pinpoint of matter no larger than the period at the end of this sentence. This was a fertilized human egg. You came from such a fertilized egg, and this was a very remarkable cell indeed. It contained all of the genes necessary, not only to make you a human being, but a human being with characteristics of a particular race and characteristics peculiar to your immediate family. This cell was formed by a union of a sperm from your father and an egg from your mother. The sperm was so small that a thousand of them could be lined up side by side across the head of a pin with room to spare. Yet this was the only hereditary link which you have with your father—within the tiny head of this sperm were all of the genes for the characteristics which you have inherited from him. A man may die before his child is born, yet that child may show great similarity to his father in his appearance, mannerisms, and aptitudes. The egg from your mother was very small also, but still many times larger than the sperm. This does not mean that you inherited more from your mother. The egg contained approximately the same number of genes as the sperm.

These two cells—the sperm and the egg—form the bridge of life which connects the generations of the past with those of today, and it is such cells which will carry the genes to the generations yet unborn. Over this slender bridge all of the hopes of the past and the prospects for the future of mankind must pass. In this chapter we shall learn how these important reproductive cells are formed and how they unite to start a new life. We shall begin with a study of the genes and how they duplicate during cell division.

TWO CELLS FROM ONE

An adult human being has literally trillions of cells in his body, and the great majority of these each contain the same number and kind of genes which were present in the one cell from which they all developed. There are a few exceptions, the red blood cells, for instance, do not have nuclei nor genes, but the cells which formed them did have genes. Body growth hinges upon growth of the cells, followed by a division, then more growth, more division, etc. Such cell divisions are not a mere splitting in two, for this would give two half-cells, cells with only half the genes found in the original. Instead the divisions must be by a special process, **mitosis,** which involves first a duplication of genes, then the chromosomes upon which they lie, and finally a segregation of the duplicated parts so that, as the cell splits in half, each of the two daughter cells receives a complete set of genes and chromosomes.

The Duplication Of Genes And Chromosomes.   The number of genes in a human cell is estimated to be as many as three hundred thousand, so they must be very small, since they are all contained in the nucleus of the cell. They are arranged in linear order on long, slender bodies known as **chromosomes.** Man has forty-six of these chromosomes in each body cell. They vary in length, the longest being about five times as long as the shortest, so there is also variation in the number of genes on each chromosome.

Most of the cells of your body are not in any stage of mitosis and are said to be in the **interphase.** If you could focus on the

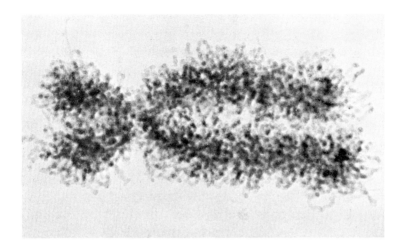

FIGURE 2-1

A human chromosome, carrier of the genes. This photograph was made through an electron microscope which has magnified the chromosome about 100,000 diameters. This chromosome is in the prophase stage and has already duplicated, but the two chromatids are held together by the centromere. The complex coiling of the gene strands shows clearly under this very high magnification. (Photograph courtesy of E. J. DuPraw)

interior of a skin cell with a very high powered microscope, you would see the chromosomes as slender threads, each of which would have a small constriction, the **centromere,** somewhere along its length. The genes are functioning, directing the activities of the cell by means of messages which they send out into the cytoplasm of the cell. Then, something happens; some sort of trigger signals the genes to duplicate.

Genes are all made of a chemical known as **DNA** (deoxyribonucleic acid). This is in the form of two long twisted strands, somewhat like two long pieces of string twisted together. Between the strands are four kinds of cross-connections. Genes vary according to the arrangement of these cross-connections.

This is considered in detail in Chapter 8. When the stimulus for duplication comes, the strands begin to split apart at one end, each strand taking half of each cross-connection with it. Then each half gene attracts to itself the missing pieces, which lie free in the cell as part of the food, and two genes have been formed from one, each an exact duplicate of the original.

Perhaps you can visualize this better if you think of a gene as a long stick of peppermint candy made of two coiled stripes, one red and one white. Then assume that the two stripes split apart and that the candy is in a solution containing small pieces of red and white candy. The red stripe attracts to itself the white pieces and forms them into a white stripe, and the white stripe forms a red stripe in the same manner. Thus, we end up with two sticks of candy, each an exact duplicate of the original stick.

There is then a period of about four hours before the cell shows any sign of mitosis. Then it enters the **prophase,** and the chromosomes become shorter and thicker by coiling somewhat like a spring. At this stage it can be seen that each chromosome is now double, made of two **chromatids,** but still having a single centromere which holds the two chromatids together.

The process of mitosis consumes about forty-five minutes during which time the chromosomes continue to shorten by continued coiling until they are relatively short rods. Then they move to the center of the cell and are engulfed by a spindle figure which is made of many small fibers. The chromosomes line up in the center of this **spindle figure** and a spindle fiber becomes attached to the centromere of each double chromosome. This stage is known as the **metaphase** of mitosis. Each centromere now duplicates itself so that each chromatid has a centromere. The spindle fibers seem to exert a pull on these centromeres so that they separate and move to opposite poles of the spindle figure pulling their chromatids behind them. As the two chromatids are pulled apart each becomes known as a chromosome. The cell is then in the **anaphase.** Once the chromosomes reach the poles of the spindle figure, the cell enters its final stage, the **telophase.** Here the chromosomes uncoil and become longer and thinner; a new nuclear membrane forms around them, and the cell begins to split in two. Thus two daughter cells are formed, each with all the genes and chromosomes that were found in the original cell.

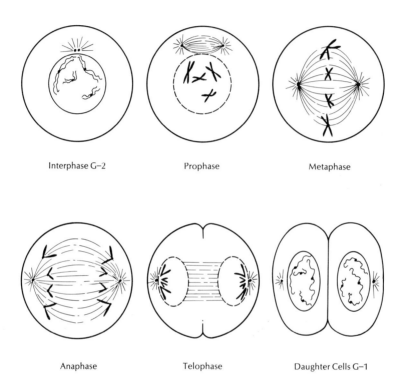

FIGURE 2-2

Cell duplication by mitosis. Only four of the forty-six chromosomes are shown here, but the events involved are the same for all. The interphase is shown in the G-2, when the chromosomes are already duplicated, while the daughter cells are shown in the G-1, when the chromosomes are single.

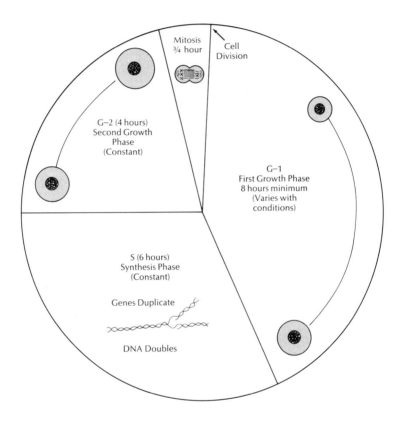

FIGURE 2–3

The cycle of cell growth and duplication under optimum growing conditions. When conditions become less favorable, the time in the G–1 will be extended, but the other parts of the cycle will remain constant.

The Timing Of Mitosis. If human cells are grown in tissue cultures with plenty of space and food, they will establish a time pattern of growth and division. After mitosis, there follows a **first growth period (G–1)** when the uncoiled chromosomes are putting out their messages and the cytoplasm is very actively producing new protoplasm. After about eight hours, the genes duplicate. This is known as the **synthesis stage (S)** which lasts for about six hours. Growth slows during this time because genes cannot put out messages while they are duplicating. Then follows a **second growth period (G–2)** of about four hours. Finally, **mitosis** takes place, requiring about forty-five minutes, and the entire cell cycle is repeated for each daughter cell. Of course, not all cells of your body duplicate at this rate; this is the minimum time. When there are longer times between duplications, the first growth period is extended. This may be twelve hours, twenty-four hours, a week, a month, or even longer. In fact some of your cells will never again go into the synthesis stage. Once the synthesis stage begins, however, the time schedule is adhered to. That is, mitosis is going to follow in about ten hours regardless, as long as the cell stays alive. Some cells, once matured, never grow nor divide again; brain cells are an example. They are said to be in the **zero growth stage (G–0).**

What determines when a cell shall enter the synthesis stage of gene duplication? This is a key question, the answer to which would solve many of the mysteries of life. Cells on your skin, for instance, may divide at an average rate of once a week, just enough to replace the cells sloughed off. However when you injure the skin, this rate is stepped up to its maximum until the injury is repaired, then the cells return to their slower rate. One theory holds that the injured cells produce a wound hormone which acts as a stimulus to start the synthesis stage.

In some cases the cells continue their maximum rate of growth and duplication when new cells are not needed; this leads to the formation of a **cancer.** Cancer research is seeking ways to turn off this uncontrolled cell cycle.

Mitosis can account for the continued production of new cells to produce and maintain a human body, but the human body cannot continue forever. Old age and death will stop the life processes of all in the course of time. Hence there must be a way to transmit the genes to other generations if human life is to continue. Sperms and eggs are special cells which have this purpose. Let us turn our attention first to the method of sperm production.

THE FORMATION OF SPERMS

If you had the power of X-ray vision and could look inside a human testis, you would see that it is made of many feet of tiny, coiled tubules, the **seminiferous tubules,** about the diameter of a coarse sewing thread. This is where the sperm are produced.

At the outer lining of these tubules, there are some very special cells which form the **germinal epithelium.** These are the only cells in a man's body which have the power to continue the species; they can produce sperms. If you could watch one cell in this tissue, you might see it divide by mitosis. One of the two daughter cells remains in place while the other begins moving to the interior of the tube. It is known as a **primary spermatocyte.** This cell will form four sperms in a series of two divisions known collectively as **meiosis.** Meiosis has only one gene duplication, one chromosome duplication, and one cen-tromere duplication for two cell divisions. This accomplishes a reduction of the gene chromosome number to one-half; only twenty-three chromosomes will be in each sperm. This is a very necessary process; if sperms and eggs were formed by mitosis, the chromosome number of man would double each genera-tion.

The primary spermatocyte has its synthesis stage as in mi-tosis, but as it goes into the prophase, a very important difference can be seen. The chromosomes are paired. Man actually has twenty-three pairs of chromosomes. Each chromosome of one kind has a mate of the same kind. There is one exception in a man. One pair of the twenty-three is unequal in size. This pair is made of a long X-chromosome and a very short Y-chromo-some. Female cells have two X-chromosomes. As you might expect, these chromosomes are related to the determination of sex and will be studied in detail in Chapter 5. Since each chromo-some is double and paired, there are actually four parts, chroma-tids, to each pair.

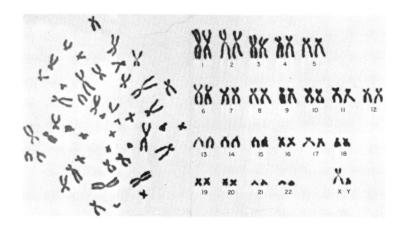

FIGURE 2–4

Human chromosomes. On the left the chromosomes are shown as they appear when a cell in prophase is squashed flat. On the right is a karyotype which has been made by cutting out each chromosome and matching them all in pairs. One of the twenty-three pairs is unequal in size. These are the X- and Y-chromosomes which are related to sex determination.

At the metaphase of this first meiosis the chromosomes are lined up at the center of the spindle, and spindle fibers pull the pairs apart by their centromeres. Two daughter cells, **secondary spermatocytes,** are thus formed, each with only twenty-three chromosomes, but each chromosome is double. A second division is thus necessary to reduce the chromosomes to a single state. There is no synthesis stage preceding this second division, so the chromosomes enter the prophase in the same condition in which they were at the telophase of the first division. At the metaphase the centromeres duplicate and pull the two chromatids of each chromosome apart, so the four cells which result have only twenty-three single chromosomes. These cells are known as **spermatids.**

Each spermatid now is transformed into a sperm. The nucleus becomes compacted into a body which will become most of the head of the sperm. At the front of the head, however,

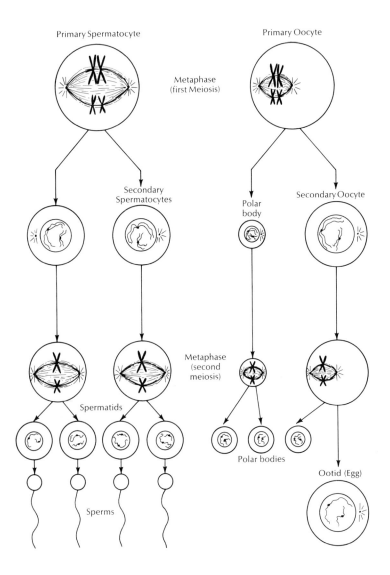

Primary Spermatocyte

Primary Oocyte

Metaphase
(first Meiosis)

Secondary
Spermatocytes

Polar
body

Secondary Oocyte

Metaphase
(second
meiosis)

Spermatids

Polar bodies

Ootid (Egg)

Sperms

FIGURE 2–5

How sperms and eggs are produced by meiosis. Compare this diagram with that of mitosis (Figure 2–2) and note how meiosis reduces the chromosome number in half during its two divisions.

bound enzymes accumulate to form the **acrosome.** These enzymes will be needed to penetrate the egg. On the other side of the nucleus small, rod-like bodies, the **mitochrondria,** accumulate to form the middle piece or neck of the sperm. These are energy-yielding bodies and will furnish the power for the sperm to swim. Some of the cell stretches out to form a long tail, the **flagellum,** which will propel the sperm. The remaining cytoplasm and the nuclear components besides the chromosomes are thrown off as little blobs of protoplasm which pass down the tail and off at the end. This leaves the sperm as a very small, "stripped down" cell which carries only the genes and the accessory parts needed to get it to the egg.

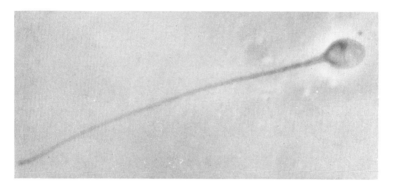

FIGURE 2–6

A living human sperm, greatly magnified. All of the hereditary potential from the father must be compacted into the tiny head of such a sperm.

Mature sperms are released inside the tubule and are carried by cilia lining the tubule up inside the body where they are stored until released.

THE FORMATION OF EGGS

When you speak of eggs, the average person probably thinks of those large products of chickens which may be eaten for breakfast. It sometimes comes as a surprise to learn that human beings produce eggs also. These human eggs are produced in a manner very similar to the production of bird eggs, only they are much smaller in size. The bird egg must be large because it must contain enough food to supply the developing embryo until it hatches and can begin feeding. Human beings, however, are mammals, and mammals have a special place, the uterus, where the embryos are nourished through an attachment to the mother. As a result, there is no need for an egg with a large amount of yolk. The egg needs only enough stored food to supply the embryo until it can make an attachment to the mother's uterus and begin absorbing food from the mother's blood.

The **ovaries** are the female reproductive glands. There are two of these in the human female, each of which is about the size of a shelled almond. They are found inside the body near the lower part of the back. As was true of the testes, the cells which are to produce reproductive cells are found in a layer of germinal epithelium around the outside of the ovaries. Cells in this tissue form **primary oocytes** which can become eggs.

In the first division of meiosis the chromosomes duplicate, pair, and separate as in sperm formation. One peculiar difference can be noted, however. The spindle figure is not across the center of the cell, but to one side. Then when the cell divides at the telophase, only a little of the cytoplasm is pinched off with one nucleus, leaving most of it with the other nucleus. This results in one large cell, the **secondary oocyte,** and one small cell, the **polar body.** The secondary oocyte undergoes a second division of meiosis and gives an ootid, actually the egg, and a second polar body. The first polar body may also undergo a second division, so we end up with one egg and three polar bodies.

The primary oocytes are formed early in embryonic life, and the first division of meiosis takes place between the fifth and seventh month after conception. Then the secondary oocyte is held at this stage until ovulation which may be from twelve to forty-five years later. This long delay can sometimes cause the chromosomes to adhere and give abnormal distribution to the egg. This is discussed in Chapter 10.

Why should the divisions of the cells be equal in sperm formation, but unequal in the formation of the egg? An egg must contain enough food to supply the developing embryo until it can establish a connection with the uterus of the mother. Hence, the egg must be relatively large to contain sufficient yolk granules to furnish this food. Four equal divisions would reduce the size accordingly. Also, there is no need for large numbers of eggs as is true of sperms. Each egg has a relatively high chance of being fertilized. An average woman will release no more than about three hundred and fifty eggs during her lifetime. This may seem large compared with the number of children which a woman can bear, but it is very small when compared with the hundreds of millions of sperms which are released from a man's body at one time.

Now we have seen how one half of a person's genes are concentrated into reproductive cells which can bridge the generations. In the next chapter we shall see how these cells come together and unite to form a cell which can produce another human being.

Chapter 3

A NEW LIFE BEGINS

One of the greatest marvels of nature is the beginning of a new human life. Two cells from two different people are brought together by a complex set of psychological and psysiological factors. Neither of these cells is capable of survival alone for very long, yet when united they form a cell with the potentialities of producing a human being. After many divisions by mitosis this cell can form an embryo which is recognizable as a human being, and with continued growth it can make a mature person with the capabilities of reproducing more human beings.

Thus, we can say that a new life begins when a male reproductive cell, the sperm, unites with a female reproductive cell, the egg. In many water animals the bringing of the two cells together is a comparatively simple process—the females lay their eggs, and males release their sperms in the water. The sperms then swim about and come in contact with the eggs. The head of the sperm, with its valuable cargo of genes, enters an egg, and fertilization is accomplished. For animals which do not live in water the process is more complicated—the sperms must be deposited within the body of the female, and fertilization takes place there. Such a method requires a complex set of organs and reactions not needed by most water animals, although there are some water animals which also have internal fertilization. The reproductive process in mammals is further complicated by the fact that the young embryo is protected and nourished in the body of the mother for a time after fertilization and is nourished by milk from the mother after birth. Man is in this latter group. In this chapter we shall explore some of the events connected with the beginning and early development of a new human life. Studies on very early embryonic development have been made by fertilizing human eggs in the laboratory and studying the development under the microscope. Information on later development has been obtained from studies of embryos which have been expelled from a woman's body at various stages before the time for normal birth.

FERTILIZATION

When a girl reaches the period of her life when she is capable of reproduction, she will begin a hormone cycle which stimulates the release of eggs from her ovaries at a rate of one about each twenty-eight days, although this time may be quite variable. As an egg is released it is caught in a tube, the **oviduct,** and is carried toward a pear-shaped organ, the **uterus** or womb. The egg is capable of being fertilized for only a relatively short time, less than a day, after it is released. If no fertilization occurs, the egg continues its way down the oviduct to the uterus and is then expelled from the body. Sperms also have a limited useful life. Just before being expelled from a man's body, they are mixed with secretions of special glands which help to carry the sperms and also activate them. The mixture of sperms and this fluid is known as **semen.** Sperms, as they are stored within the body, are relatively inactive. They have only a very small food supply, and this must be conserved to be used to carry them to the egg. Once they are mixed with the secretions, however, they become quite active. When viewed under the microscope, their tails can be seen threshing about vigorously as they swim. Such a rapid expenditure of energy soon exhausts their meager food supply, and within a few hours their movements become feeble and then stop. They have lost their power to fertilize an egg.

It is possible to prolong the useful life of sperms by refrigeration. The sperms cease their movements if kept cold, and they will retain the power of fertilization for a week or longer. This method has been used by cattlemen for many years to preserve the semen from bulls, so it can be used later to fertilize cows. More recently, the semen has been frozen and in this state seems to keep almost indefinitely. Many calves have been born which were produced from fertilizations from semen which had been frozen for over a year. The method has even been used for some human fertilizations. It has been suggested that this method could be used for men who might be exposed to conditions conducive to harmful changes within the sperms. Some astronauts, for instance, who are to be in outer space for long periods of time will be exposed to high-energy radiation which can cause changes in the genes and chromosomes which might be harmful to children formed from such sperms. If semen from these men was frozen before their journey, it could be used later so they could have children from sperms which remained safely in a deep freeze here on earth. Some have even gone so far as to suggest that we establish sperm banks which could be preserved for many years. Then perhaps some men who showed highly desirable genetic qualities, as evidenced by their lives, could father children long after death.

The semen is deposited in the female's body near the entrance to the uterus. As they swim about, many sperms will enter the uterus and continue on up the oviducts (fallopian tubes) where they may meet an egg on its way down. What causes the sperms to take this pathway? Studies of sperms under the microscope show that they tend to swim against the direction of the flow of a fluid. If you create a current moving to the right, the sperms will swim to the left. In the female reproductive tract there are cilia which create a movement of a mucous secretion down the fallopian tubes, down the uterus, and on down the vagina. This movement brings the egg down the tubes from the ovary, but it stimulates the sperms to swim into the uterus, up the uterus, and up the tubes. Thus, with opposite reactions the two tend to meet.

Once they reach the egg, the sperms find a barrier to fertilization. The egg is surrounded by a corona of small cells held together by a cement. Here is where the enzymes in the acro-

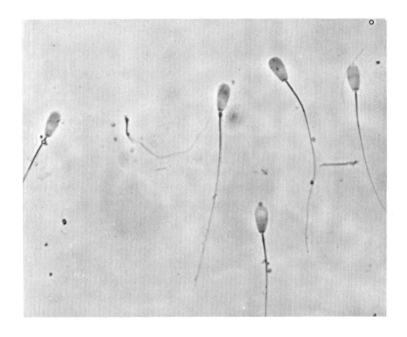

FIGURE 3-1

Orientation of sperms by direction of fluid move-
ment. The fluid was moving towards the bottom of
the field, and these sperm have responded by swim-
ming in the opposite direction. Such a response leads
the sperm up the female genital tract to the egg.

somes of the sperms are needed. Each sperm contributes its enzymes which can dissolve the cement holding the corona cells
together. Through the combined action of enzymes from many
sperms, the barrier is eventually breached, and sperms can penetrate to the surface membrane of the egg. A man will be of low
fertility if the concentration of sperms in his semen is low.
Enough sperms may not reach the egg to break through the
corona.

The egg produces a secretion, **fertilizin,** which traps the
sperm head once it makes contact. This secretion is specific;
sperms of one species generally will not respond to the fertilizin
of another species. You could not produce a cat-rabbit hybrid,

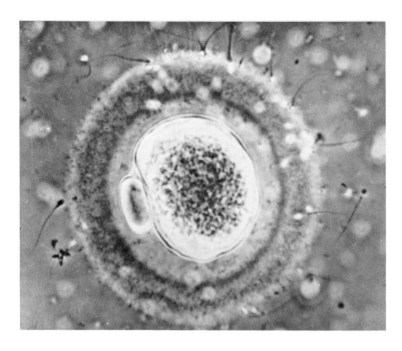

FIGURE 3-2

Fertilization, the union of sperm and egg. Sperms can be seen penetrating the corona surrounding the egg. When one contacts the egg surface, the head will be taken inside, and the potentialities of a new life will be formed. Note the polar body to the left of the egg. (Photograph courtesy of Landrum Shettles)

even by artificial insemination. The fertilizin of one would not trap sperms of the other. In the case of closely related species, however, the fertilizin is similar enough to allow fertilization. Domestic cattle and the American buffalo, or bison, can be successfully crossed, although the hybrid is sterile. The same is true of the horse-donkey cross which produces a mule.

The outer membrane of the egg bulges out around the head and middle-piece of the sperm and pulls them inside, leaving the tail behind. This entry stimulates some chain reaction which causes the egg to reject other sperms, and any other sperms which may already be stuck to the membrane are released. Thus, only one sperm head normally enters the egg.

Fertilization has now been accomplished—the sperm head has brought in twenty-three chromosomes which match the twenty-three chromosomes already in the egg. The fertilized egg now has forty-six chromosomes and the potentialities of producing a human being.

EARLY DEVELOPMENT OF THE EMBRYO

When a sperm enters an egg, it provides the stimulus which is necessary to set in motion the events leading to mitosis. The unfertilized egg has a single complete set of genes and theoretically could produce a human being without fertilization, but normally will not start its development without the stimulus of the entrance of a sperm head. In some studies on experimental animals the egg can be induced to start development by pricking it with a needle or treating it with certain chemicals. There is even one breed of white turkeys in Maryland in which many of the eggs will hatch without fertilization or without any other kind of treatment. Under normal conditions, however, fertilization must furnish the stimulus for division of the egg.

Many changes take place within the egg when it is fertilized. The rate of oxygen consumption increases greatly, to about five times its rate before fertilization. This shows that great activity is taking place within the cell. There is an increase in the permeability of the outer cell membrane, which allows substances to diffuse in and out more rapidly. Also, the contents of the cell become more viscid (thicker in consistency)—this seems to be necessary for the beginning of cell division.

Within about thirty hours after fertilization the first division has been completed, and there are two cells. Soon these both divide to give four, these divide to give eight, and so on. Three days after fertilization there are thirty-two cells clustered together to form a small ball, known as the **morula.** There is no growth of cells between divisions, so the entire mass of the embryo is no larger than was the fertilized egg. About a week after fertilization the cells begin to show some differentiation as they shift their positions and begin to assume different functions. If the embryo is cut in two at this time, it will show an outer layer of cells surrounding a hollow cavity, but within the cavity is an inner mass of cells attached to one side of the outer cell layer. The embryo is now known as a **blastocyst.**

IMPLANTATION

The embryo now faces a critical period of its existence. As it comes from the fallopian tube into the uterus, it must form an attachment to the walls of the uterus or it will pass on out of the body. Many fertilizations do not go past the blastocyst stage because they fail to achieve implantation in the wall of the uterus. One study indicated that about half of all the blastocysts fail to achieve implantation. This is good because most of these were shown to be defective blastocysts—they had not developed properly, and if they had become implanted, they would have formed abnormal and deformed babies.

As a part of the female reproductive cycle, hormones have been produced which cause the wall of the uterus to become swollen and filled with a rich blood supply. It is sticky with mucus, and when the blastocyst contacts the wall, it tends to stick to it. A normal blastocyst then has the power to produce enzymes which digest the surrounding area, and soon the embryo becomes buried in the wall of the uterus. This implantation is completed by about the tenth or eleventh day after fertilization.

The outer layer of the blastocyst now sends fine finger-like projections down into the wall of the uterus and begins to absorb nourishment from the blood of the mother. Growth and differentiation now become very rapid. The genes within the cell direct the formation of different layers of cells which produce different body organs. Some of the cells form membranes around the embryo. One membrane, the **amnion,** completely surrounds the embryo, and it is filled with a fluid in which the embryo floats. An embryo is very soft and would be easily injured during its development, but it is well protected as it floats in this fluid. Another membrane, the **placenta,** forms a close connection with the uterus of the mother and absorbs food and oxygen from the blood of the mother which is circulating through the uterus. An *umbilical cord* connects the embryo with the placenta.

THE EMBRYO'S TIMETABLE

When only four weeks old the embryo has a heart which is pumping blood out through blood vessels. The heart in proportion to the body is nine times as large as in the adult. It must be large because it must force blood not only through the body of the embryo, but also out into the umbilical cord, through the placenta, and back to the body. In size the embryo is only about a tenth of an inch long, but it shows the basic organization of a human being. At five weeks the embryo has grown to about a third of an inch long, small stumps which are to make arms and legs have appeared, and the eyes are visible as black circles. Strange to say, there is also a clearly-developed tail, which normally disappears before birth, but there are exceptional cases where the tail continues to develop along with the rest of the body, and the baby is born with a tail several inches long.

At eight weeks of age the embryo is about an inch long and is recognizable as a human being. The eyes are well developed, but the lids are beginning to stick together and will seal the eyes shut during the rest of development as a protection against possible injury. In some animals, such as dogs and cats, the eyes are still sealed at birth, but in human beings they normally open shortly before birth. There is also a beginning of bone development in the tiny skeleton at this stage of life. At eight weeks it is customary to begin calling the developing body a **fetus** instead of an embryo, although biologically it is an embryo until birth.

At twelve weeks the fetus is about three inches long and begins moving its body about so vigorously that the mother begins to feel the movements in her body. If born at this time, the fetus will respond if tickled, but it cannot survive more than a few hours. At sixteen weeks it can move its lips and swallow, clench its fists, and the face begins to have human features. At eighteen weeks the fetus is over six inches long and is quite active. It may suck its thumb and inhale and exhale, but it can

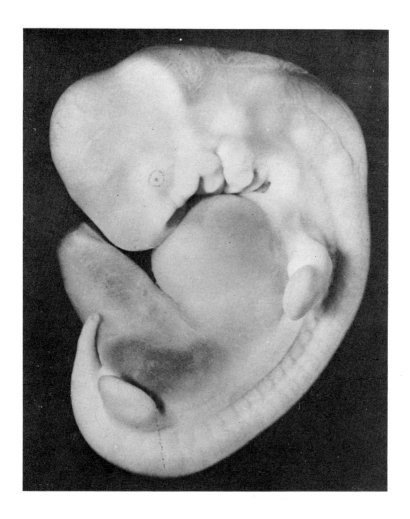

FIGURE 3-3

Human embryo thirty-two days after fertilization. The genes have used the yolk stored in the egg first, and then the food obtained through its connection with the mother to construct the basic parts of a human body. (Photograph courtesy of C. F. Reather, Carnegie Institute of Washington)

only take in the fluid in which it is floating, since there is no air present.

At twenty-eight weeks an important milestone is reached. The fetus now has a chance to live if born prematurely. It is only about ten inches long and weighs only a little over two pounds, but the body parts are sufficiently developed that they can take over independently if the fetus is placed in an incubator and given very special care after it is born. From this time on until forty weeks, which is the normal time for birth, the fetus increases greatly in size and strength and is much better able to meet the problems of life outside the uterus of the mother if it can remain for this full term of development.

Interrupting The Timetable.   We have seen that there is a specific time for each event in embryological development. Unless development proceeds according to this schedule, a normal embryo will not be formed. During the years 1961 and 1962 the world was shocked by the reports of the birth of many babies with only stumps of arms or legs in Europe. In an effort to learn why there was such a sudden large increase in such abnormalities, the physicians found that in nearly every case the mother had taken a new drug, thalidomide, during early pregnancy. Tests on animals and human beings seemed to show no harmful effects, so it was prescribed to relieve some of the unpleasant symptoms of nausea which frequently accompany early pregnancy. Some women who used it reported a temporary numbness of their arms and legs, but this soon passed away, and there was no permanent damage. For young embryos in their bodies, however, it was a different story. The arms and legs might be slowed in development at a critical period of their timetable. Once this time was passed, the limb buds did not go back and do what was supposed to be done at this period. Instead, this part of the development was just skipped. As a result some of the babies were born with arms or legs which were somewhat like those of the early embryo. Since they roughly resembled the flippers of seals, the condition was called **phocomelia** which means seal limbs. Some were born without any arms or legs, others had only one such limb, and still others had two or three so affected.

Experiments on lower animals have turned up quite a number of such **teratogenic agents.** In mice and some other animals a high percentage of cleft palate was found in the offspring of females who were given the hormone, cortisone, during early pregnancy. This hormone is widely used to treat arthritis, allergies, and other difficulties in human beings. While we have no direct evidence that this hormone has a teratogenic effect on human embryos, the experimental results on lower animals should be sufficient reasons to cause us to use this and other very active hormones with great caution during pregnancy. Also, a number of other powerful drugs have come under suspicion. The number of women who have borne defective babies after using LSD during early pregnancy makes this drug highly suspect.

It is also possible for some diseases to cause abnormalities. German measles, rubella, is a virus disease which is very mild in its effect on a child, but if a woman takes it while she is carrying a young embryo, serious damage can result. The virus can penetrate the placenta and infect the developing embryo. The most common abnormalities which result are eye defects (especially cataract which causes blindness), heart defects, deafness, defective teeth, small head and brain with idiocy, and mental retardation. The nature of the defect depends to some extent on the stage of the embryo when the infection develops. Cataracts of the eyes tend to form if the infection is in the sixth week, deafness if it is in the ninth week, and heart defects if it is in the fifth week. We now have a vaccine for rubella, and public health authorities are trying to give it to all young school children. This is not because the disease would harm the children if they caught it, but because they may take the virus home where it could infect their mothers who might be pregnant. Other virus diseases which can cause trouble are hepatitis, mononucleosis, and *Herpes zoster* (shingles). Even influenza is suspected.

Public health officials have long known that the frequency of new born abnormalities varies, showing peaks and valleys which are not related to seasons. Some studies made in Denver indicate that the high frequencies follow outbreaks of virus diseases by about six to eight months. Some physicians have felt that the danger is so great that they will advise a woman to have a therapeutic abortion if she has such a disease during early pregnancy.

DETECTING ABNORMALITIES BEFORE BIRTH

As liberalized abortion laws have made possible the termination of a pregnancy when an abnormal embryo might be born, doctors had hopes for some discovery which would enable them to know for sure whether an early embryo was going to express a serious abnormality. Now such a method is available. The amniotic fluid in which the embryo floats contains many cells which have broken off from the skin and other body parts. By a technique, known as **amniocentesis,** a slender needle can be inserted through the mother's abdomen and a sample of this fluid removed. This can safely be done as early as twelve weeks after conception. The cells in this fluid can then be grown in culture dishes and studied.

Tests can be made to see if these cells have the power to bring about normal chemical changes. An inability to make one such change may reveal the absence of a vital enzyme, without which the embryo cannot be normal. At present over forty such abnormalities can be detected by this technique, and the list is growing all the time.

As an example, consider a Philadelphia couple who bore a beautiful child who was normal and healthy. At four months, however, the baby gradually became more and more lethargic, and its mental abilities began deteriorating. Doctors' diagnosed the condition as the dread **Tay-Sach's disease.** The baby could not make one of the cell enzymes needed to process fats, so fatty material was accumulating in its nervous system and was gradually destroying the nerve cells. Death would result within two or three years. Both parents carried a recessive gene for this condition, and their first child happened to get the gene from each parent. Now they were greatly concerned about future children since each child would have a 25 percent chance of having this same abnormality. With the assurance from their doctor that it could be recognized, they began another pregnancy and at about twelve weeks amniocentesis was performed. A test of the cells floating in the amniotic fluid showed that they could process fats properly, hence they knew it would not have the disease. Not only were they relieved of a great worry, but had the child been abnormal they had decided to terminate the development while the embryo was only about three inches long, rather than wait for a horrible death after birth. Of course, a moral issue is at stake here. Some feel that man has no right to terminate a life, even though it is only a partially developed embryo unable to survive outside its uterine environment, but modern laws make this an issue to be decided by each set of parents.

It is also possible to see the chromosomes in cells obtained from the amniotic fluid and recognize many abnormalities resulting from abnormal chromosomes. Some of these are described in Chapter 10.

TWO OR MORE BABIES AT ONCE

About one percent of the women in the United States who go to a hospital for a delivery come home with two babies. This means that about one person in fifty is a member of a twin pair. In much rarer instances, there may be three, four, or even five babies born at the same time. Why does the reproductive process sometimes result in multiple births? It is normal in many other mammals, but unusual in human beings.

First let us consider the two types of multiple births. You have surely seen many twin pairs, some of which looked so much alike that you may have had difficulty distinguishing one from the other. Still other twins show many differences; they are no more alike than brothers or sisters born at different times, in fact they may even be of different sexes. The first type is known as identical or monozygotic twins. They both originated from one fertilized egg or zygote. The second type is known as **fraternal** or **dizygotic** since they originate from two different fertilized eggs. Let us consider the first type.

Thanks to the exactness of mitosis in giving each daughter cell a complete set of genes, each of the two cells formed by the first mitosis of the zygote has the power to produce a complete person. Normally these two stay together, and each cell forms only half a person, but should they fall apart each cell will produce a complete person. These two will have exactly the same genes, so will be identical or monozygotic twins. They will be of the same sex, and all inherited traits will be the same. Sometimes the splitting comes later, after the embryo is a ball of cells, but the results will be the same. If it splits into two halves before becoming implanted in the wall of the uterus, each will be separately implanted and each will have its own complete set of membranes. These are **dichorionic twins,** a name taken from the chorion, the middle membrane of the three around the embryo. If implantation takes place first, something may go wrong with the gene correlation, and a second embryo will bud off from the first. These will both be inclosed in the same chorion and are **monochorionic twins.** If the budding is rather late, they may even both share the same amnion. About 70 percent of the identical twins in the United States are of this monochorionic

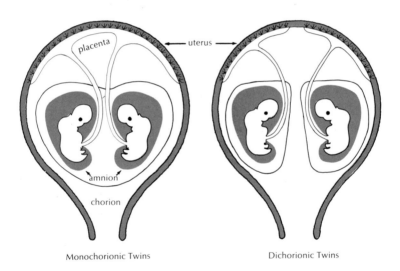

Monochorionic Twins   Dichorionic Twins

FIGURE 3–4

The two types of twins according to embryonic membranes. Only identical twins share the same chorion, but both types of twins may be dichorionic. Sometimes the monochorionic twins also share the same amnion.

type. Occasionally the splitting is not complete, and we have conjoined or **Siamese twins** with varying degrees of junction.

The dizygotic twins are due to double ovulation. Normally as a woman ovulates, a hormone is released along with the egg and shuts off the hormone from the pituitary of the brain which stimulates ovulation. At times, however, a second ovulation takes place before the shut off is accomplished, and two eggs are available for fertilization at the same time. All of the fraternal twins will have separate implantations and will be dichorionic.

The splitting of an embryo to form identical twins seems to be an accident which is rather uniform in all races and not influenced by heredity. Multiple ovulation to give fraternal twins, however, varies greatly in different races and in different families, so heredity must be involved. We can calculate the percentage of each by determining how many twins are of opposite sexes. In Caucasian whites in the United States this is about 31 percent. Since half of all fraternal twins would be expected to be of opposite sexes, then there must be another 31 percent of the same sex, giving a total of 62 percent. This leaves 38 percent identical twins. Studies of other races show monozygotic twins to be constant at about three per one thousand births. For dizygotic twins, the following variable figures have been found:

TABLE 3–1

Race	Dizygotic twins per thousand births
Amer. Caucasian	6.6
Japanese	2.5
Nigerians	39.9

Heredity influences hormone production, and it is hormonal variations which cause multiple ovulation.

Man can manipulate the hormones and induce ovulation. Sometimes women are sterile or of low fertility because they do not ovulate regularly; there is a deficiency of the hormone which stimulates ovulation. Many such women have been able to bear children by taking the hormone which stimulates ovulation. The main problem is that it is difficult to know just how much hormone to inject. Too much will result in multiple ovulation. We have many cases of multiple births coming to women who had gone for years without becoming pregnant. A couple in New Jersey recently had five children at once after the wife took this hormone.

The birth control pill works in just the opposite manner. It inhibits the hormone which stimulates ovulation, so there is no ovulation and no pregnancy. This causes a storing up of eggs ready for ovulation and when the pill is suddenly stopped, there may be multiple ovulation. Some women are advised to stop the pill and use other means of birth control for several months before becoming pregnant to avoid the chance of multiple births.

Other interesting facts have been uncovered by statistical studies of twin frequencies. A change from warm to cold weather is conducive to multiple ovulation. In Texas, physicians have noted that there is a sudden increase in twins about nine months after a chilling Texas norther hits the state.

Multiple births greater than twins may be either monozygotic, multizygotic, or even a mixture of the two. The famous Dionne quintuplets of Canada all started as one fertilized egg. This split itself into a number of parts, five of which became implanted and the result was five baby girls all identical. The Keyes quadruplets of Texas started as three fertilized eggs and then one of these made two. The New Jersey quintuplets were all due to separate zygotes and were of mixed sexes.

The study of twins and other multiple births can give us much valuable information on the relative influence of heredity and environment. Identical twins have identical genes and any differences they show must be due to environment. Fraternals, on the other hand, have some differences in genes as well as environment so they can express differences from both causes. This important topic is explored more fully in Chapter 11.

Chapter 4

OUR GAMBLING GENES

Life is a game of chance—very often chance happenings have a great influence on your life. A chance meeting may determine the person with whom you will share life in marriage. A chance contact may be the deciding factor in your choice of a vocation. Chance greatly influences where you will live and your religious affiliation or lack of one. Certainly life is a gamble because so much of its course is influenced by chance happenings.

The most important gamble of all, however, took place before you even existed as an individual, when the genes were being assorted in the reproductive cells of your parents. You can only receive half of the genes of each parent, and chance determines which genes these shall be. During sperm formation the chromosomes in your father were assorted and segregated so that a particular combination of twenty-three resulted. None of your brothers or sisters received this exact assortment. The same thing happened in the production of the egg in your mother. Then, chance determined that these particular reproductive cells should come together out of the many millions of sperms and the many possible eggs which could have united.

This particular gene combination gave you all the potentials that you have ever had or will ever have. It is a unique combination, and never again can this particular combination of genes come together. Only if you have an identical twin do you have to share your gene combination with any other person. In this chapter we shall learn something about these genes and how they are assorted as they are passed down through the generations.

DOMINANT AND RECESSIVE GENES

Without an understanding of the nature of genes and how they are assorted, heredity may appear to be strange and unpredictable. A man with red hair marries a woman with brown hair and all children have hair like their mother. When these children marry, however, some of their children have the same carrot-topped condition of their grandfather. What happened to the genes for red hair in the first generation, and why were they expressed in the second generation? Another couple has two normal children, then they have one that is extremely retarded mentally. They are told that the retardation is inherited, but how could it be inherited when no one in the family, as far back as they can trace, has any sign of such a trait? These and many other questions can be answered with an understanding of the different kinds of genes.

Genes exist in pairs, known as **alleles.** For each gene you have which influences the formation of the brown pigment, **melanin,** in your skin, hair, and iris of your eyes, there is an allelic gene which also influences this synthesis. These alleles lie exactly opposite one another on the paired chromosomes in the prophase and metaphase of the first division of meiosis. They may differ in their affect; one gene of the pair might produce the enzyme needed to promote the chemical conversion of a precursion substance into melanin. The allele of this gene might not produce this enzyme. In such a case you would have normal pigmentation of your skin because one gene can produce all the enzymes you need. Hence, we can say that the gene for pigment formation is dominant over its recessive allele. The recessive gene will be expressed only when no dominant allele is present, and this is usually when both members of the allelic pair are of this recessive type. A person who expresses this gene will be an **albino;** he will have no melanin in his skin, hair, or the

FIGURE 4–1

Albinism, a recessive trait. This brother and sister had normal parents, but both parents carried the recessive gene. Three other children were normal because they received at least one dominant allele for normal pigmentation. The parents were first cousins, and this fact increased the chance that both would carry the same recessive gene.

iris of his eyes. Hence, his skin will be very fair and pink because of the blood showing through; the hair will be almost white, and the eyes a pale blue or pink.

Not many people are albinos, only about one in twenty thousand in the United States, yet about one in each seventy-one carry the gene along with its dominant allele. Hence, this is a comparatively rare recessive gene; others are more common. About one person in fifty carries the recessive gene for **PKU** (phenylketonuria) which leads to mental deterioration. About one in twenty carries the gene for **cystic fibrosis.** Some recessive genes may be found in most persons. Blue eyes can only be present when a person has two of the genes, yet in some areas of northern Europe most of the people have blue eyes. It is the frequency of a gene in the population rather than its dominance or recessiveness which determines how many people show the trait.

We need to learn a few of the geneticist's terms in order to intelligently discuss heredity. When both members of a gene pair are alike, a person is said to be **homozygous** for that gene. This could be either two genes for normal pigmentation or two genes for albinism. A **heterozygous** person, on the other hand, would have one gene of each kind of an allelic pair. We speak of the **genotype** of a person as his type of genes, and **phenotype** as the trait influenced by these genes which he expresses.

Also, we need to understand the shorthand used to represent genes. Generally letters are selected; the first letter of the trait which deviates from normal is the usual choice. For instance we use a to represent the gene for albinism and A for the dominant allele. But we use H for the dominant gene causing **Huntington's chorea** and h for its normal recessive allele. The dominant allele causes a person to lose control of his muscles at about thirty years of age.

The diagram in Figure 4–2 shows how genes are assorted and the proportions of the different kinds of offspring which might be expected. We must keep in mind, however, the fact that these are probabilities, and that every pair of heterozygous parents with four children will not have exactly three normal and one albino. In one case a young couple had just had an albino child as their first born. When the principles of the inheritance were explained along with the ratio of expected albinos, the mother seemed greatly relieved. She said, "I am so glad to know that—now that we have had the albino, we can have three more children who will be normal." Unfortunately, this is not true. Each child results from the chance union of a sperm and an egg, and half of these reproductive cells carry the gene for albinism. These cells do not know that the couple has already had an albino, and two cells carrying the gene for albinism are just as likely to unite on the second pregnancy as they were on the first. We should think of genetic ratios as the chance of each child showing the traits involved.

INTERMEDIATE AND CO-DOMINANT GENE EXPRESSION

Allelic genes do not always have a simple dominant-recessive relationship. Sometimes both genes may be partially expressed;

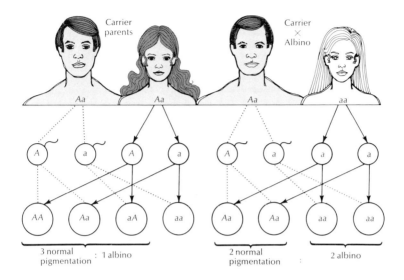

Carrier parents

Carrier × Albino

Aa   Aa   Aa   aa

A   a   A   a   A   a   a   a

AA   Aa   aA   aa   Aa   Aa   aa   aa

3 normal pigmentation : 1 albino   2 normal pigmentation : 2 albino

FIGURE 4–2

*Two kinds of genetic ratios. When normally pig-
mented parents both carry the recessive gene for albinism, an average of one-fourth of their children will
receive two of these genes and will be albinos. When
one parent is a normal carrier and the other is an
albino, the chance of an albino child is one-half.*

both may be fully expressed, or one may be expressed primarily
while the other is expressed only to a small degree.

Intermediate inheritance is clearly demonstrated in many
crosses of lower animals. In short-horned cattle, a red bull
crossed with a white cow yields offspring which are roan, a shade
in between red and white. The heterozygous animals show partial expression of both the gene for red and the gene for white.
A cross between two roans will have an offspring ratio of 1 red :
2 roan : 1 white.

A number of cases of such inheritance in man have been
found. One gene is needed in the production of normal hemoglobin, that vital oxygen-carrying component of the red blood

cells. An allele of this gene alters the structure of the hemoglobin molecule slightly, and as a result the molecules tend to aggregate in long chains, especially when the oxygen level of the blood drops. These chains cause distortion of the blood cells; they become sickle-shaped, and do not carry oxygen readily. A person with two of these genes has so much sickling that he is said to have **sickle-cell anemia.** This is a very severe anemia and is

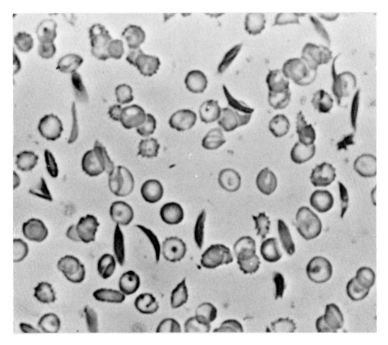

FIGURE 4–3

Blood from a person with sickle-cell anemia, a trait with intermediate inheritance. This person is homozygous for the gene and has many red blood cells formed into sickles. A heterozygous person will have some sickling, but usually not enough to cause severe anemia.

likely to lead to early death. A heterozygous person will be in between; some of the hemoglobin will be normal, and some will be of the abnormal type. As a result, he will at times suffer from a mild anemia, but mostly he can lead a normal life. He must be careful of conditions of low oxygen level, such as high altitudes because these stimulate sickling of his blood cells. He has the **sickle-cell trait.**

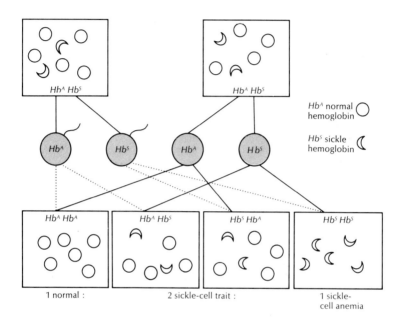

FIGURE 4-4

Intermediate inheritance of the sickle type of hemoglobin. When both parents have the sickle-cell trait (are heterozygous), their children will tend to fall into a ratio of 1:2:1.

Choosing letter symbols to represent genes which have an intermediate expression is somewhat difficult. We cannot use the capital and the small letters because they indicate the dominant-recessive relationship. Still we must use the same basic letter to indicate that the genes are alleles. The problem is solved by using superscripts. We can use Hb as the symbol for hemoglobin and add superscripts to that. For instance, the gene for normal adult hemoglobin is Hb^A, while that for the allele is Hb^S.

Skin color in man shows typical intermediate inheritance. A marriage between a very dark-skinned person and one with fair skin will result in children who are about half way in between. A marriage between persons with such intermediate shades of skin will result in a segregation of genes for dark and fair, so some children will have dark skins; some will have fair skins, and some will be in between. There are more than a single pair of alleles involved, so there will be a number of different shades in between, as will be explained later in this chapter.

The genes involved in producing the A and B antigens which result in blood types show typical **co-dominant inheritance.** When a person who is a homozygous type A marries one who is a homozygous type B, all of their children will be heterozygous for the two alleles. Their blood types will be AB. Both genes are fully expressed, so they are co-dominant.

RECOGNITION OF CARRIES OF RECESSIVE GENES

There have been many cases of genes which seemed to be entirely recessive, but upon careful investigation it was found that heterozygous carries of these genes expressed some small effect of the recessive gene. The gene for cystic fibrosis, for instance, is considered to be recessive, yet normal carriers of the gene can be recognized. Those homozygous for the gene have a defect in the utilization of salt, sodium chloride, and this results in abnormal mucous secretions. The mucus is very sticky and tends to clog up the ducts of the pancreas, the lungs and other mucous secreting areas. It is a very serious affliction with death coming in childhood in most cases. About one normal person in twenty carries the gene. Close relatives of those who have had cystic fibrosis have a much greater chance of being carriers. When two carriers marry, there is a one-fourth chance that any child will have cystic fibrosis. Now we can recognize the carriers. Analysis of the perspiration or clippings from their nails will reveal a slightly higher than normal amount of sodium in carriers. Thus, forewarned, a couple found to be both heterozygous must decide if this is too great a chance for the little gamblers who will be their children. They might feel that adoption would be a better way to have a family. With modern techniques available they could start a pregnancy, then have amniocentesis to determine if the embryo will have cystic fibrosis, and therapeutic abortion if the test is positive.

GENES THAT KILL

Quite a number of genes which result in deviations from the most common or normal state results in a lowering of viability. Some who express the trait may die or be handicapped. There are variations in the degree of reduction of viability according to the seriousness of the deviation. The most extreme genes in this series cause death in practically all who express them. These are known as **lethal genes,** genes that kill. The time when they cause death varies according to the type of abnormality. Some lethal genes may cause an abnormal formation of the cells of the very young embryo, so that it dies before it is implanted in the uterus. A woman would never know that an egg had been fertilized when this type of lethal is expressed. Other lethal genes express themselves at various times after implantation. If the gene causes defective heart development, death would come at about four weeks of age because it is at this time that the heart must begin pumping the blood if the embryo is to survive. At the other extreme, if a gene prevents normal kidney development, the embryo can live through a normal term because the mother's blood removes the waste from the embryo's blood before birth. At birth, however, the baby loses this connection with the mother and, if the kidneys cannot properly remove the waste, the baby will die.

The human body is such a complicated structure that there are thousands of possible lethal genes affecting different vital organs. It has been calculated that each person carries an average of about four lethal genes, genes which would kill had they been homozygous. When persons who are closely related marry, such as first cousins, there is a much greater likelihood that both will carry the same lethal genes because they have common ancestors. As a result, there are more lethal genes expressed, as well as other harmful recessive genes, in the children of such marriages.

This is easy to explain. Suppose a recessive lethal gene causing lung adhesions, which prevent the newborn from breathing, occurs in one person out of a hundred. This means that the chance of two people marrying, where both carry the gene is one out of ten thousand (1/100 × 1/100 = 1/10,000). A fourth of their children will be expected to receive both recessive genes, so we end up with only a chance of one in forty thousand that the trait will be expressed in random marriages. Should a man marry his first cousin, however, the chance is much greater. His chance of carrying the gene is one in a hundred. If he does carry it, the chance that his cousin will have the same gene is one in eight because of their common ancestry. Thus, the chance of both having it jumps to one in eight hundred and, with a fourth of the children expressing the gene, the chance will be one in thirty-two hundred. Thus, first cousin marriages will be expected to have about twelve times as many babies expressing this one lethal gene than non-related marriages. Table 4–1 shows the chance for some other traits. Note that the more rare the gene, the higher will be the proportion of expression from cousin marriages.

Most lethal genes which people carry are recessive; a dominant lethal may arise as a mutation, but it will kill the first person who receives it, so it is self-eliminating. It is possible, however, for people to carry lethal genes which have some intermediate expression when they are heterozygous. One such gene causes the fingers to be extremely short when it is heterozygous. The middle joint of the fingers is so greatly reduced in size that it appears as though the person has all thumbs. We say he has **brachyphalangy.** In a marriage of two people with these short fingers it has been found that about one-fourth of the children are normal, about one-half have the short fingers, and about one-fourth are born without any fingers or toes and with other defects of the bones of the body which cause death shortly after birth. This is the typical ratio of intermediate genes. It is quite probable that rare human abnormalities that we call dominant are actually of the same nature. We have never had an opportunity to know for sure because there have never been marriages of two people who both had the same such abnormality.

TABLE 4-1

Chance of Expression of Recessive Traits
in First Cousin Marriages

Trait	Frequency from Non-related Marriages	Frequency from Cousin Marriages	Ratio of Increase
Albinism	one in 20,000	one in 2272	8.1:1
PKU (phenylketonuria)	one in 10,000	one in 1600	6.2:1
Cystic fibrosis	one in 1600	one in 640	2.1:1

INHERITANCE OF CHARACTERISTICS
WHICH DIFFER IN DEGREE

As you have been reading about the methods of human inheritance, you have probably realized that all traits do not fall into simple dominant-recessive, intermediate, or co-dominant types. Take body height, for instance—we cannot group all people into just short or tall as would be expected if a dominant and recessive gene were involved. Neither could we group them into just short, medium, and tall, as would be expected if an intermediate type of inheritance was involved. There is a wide variation in degree of height which can be explained if we assume that not one, but many pairs of genes are involved in the expression of this trait. Such traits are said to have **polygenic inheritance.**

Of course environment plays a part in body stature—the food you have eaten, the diseases you have had, and other factors in your past environment have had some influence on your present body stature. Heredity, however, provides the potentialities, and in any group of people who have had a similar background the genes are the predominant force in determining stature. To illustrate, let us assume that there are four pairs of genes involved in the growth of the body in stature and that all four of these are intermediate in their effects. If two medium-sized persons marry, each carrying four genes for tallness and four for shortness, they could have children of nine different sizes so far as heredity is concerned.

Inheritance of skin color is another human characteristic which can vary considerably as a result of variation in a number of different pairs of genes. Environment again plays its part—

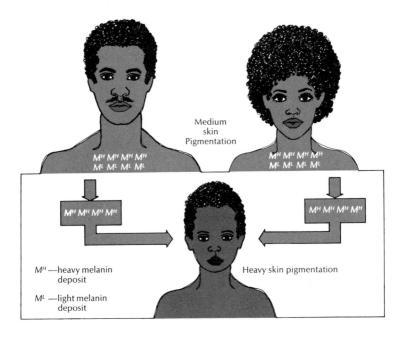

FIGURE 4–5

Polygenic inheritance of intensity of skin pigmentation. A number of genes are involved, all intermediate. Each gene for heavy melanin deposit makes the skin a little darker. In this illustration, all the genes for a heavy deposit of melanin have been received from both parents by one child. Hence, he has a much heavier deposit than either parent.

exposure to sunlight tends to produce a darkening of the skin, so when we are studying the effects of heredity, we must consider those parts of the skin which are not ordinarily exposed to sunlight. Considering modern bathing suits, only relatively small areas of the body would fall into this category. The continuous range of variations in the degree of skin pigmentation can be accounted for by polygenic inheritance of this trait.

Sometimes the expression of a trait may depend upon several different pairs of alleles, yet a variation in just one of these gene pairs can cancel out the combined effects of all the others. For instance, a person may inherit genes for a heavy deposit of melanin in his skin, yet if he also inherits a pair of recessive genes for albinism, he will be an albino. The gene for albinism fails to produce an enzyme needed to synthesize melanin, and without melanin none of the genes for intensity of melanin deposit can have any influence. We say that the gene for albinism is **epistatic** to all of the genes for intensity of melanin deposit.

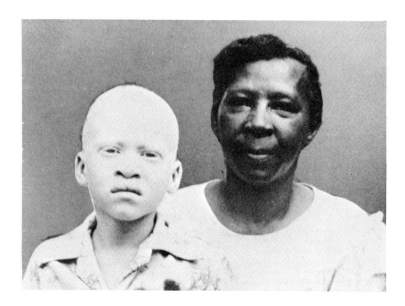

FIGURE 4–6

The gene for albinism is epistatic to genes for intensity of melanin deposit. Although this boy is from parents with heavy deposits, as shown by his mother, he received a gene from both parents for albinism. Albinism blocks the expression of his genes for heavy melanin deposit.

A midget is a very small person, usually born of normal parents who both carry a recessive gene for extreme limitation on growth. This gene might fail to produce some vital product needed in the normal output of growth hormone from the pituitary gland. The midget may actually inherit genes for a tall stature, but the one homozygous recessive gene is epistatic to all of these genes. The genes for tall stature are not altered, however, as can be ascertained by studying children of midgets. Let us say the midget is a man of about three feet tall and marries a normal woman of average height, about five feet, five inches. They may have children well above average height because of the genes for tallness inherited from their midget father.

The potential for intellectual development is a highly variable trait due to polygenes, the expression of which is highly dependent upon the environment. A couple of brilliant intellectual achievements may have a child on the idiot level of achievement in spite of maximum training. This child will have inherited a recessive gene from both parents which cancels out all the potentialities and prevents the development of a normal brain no matter what is done.

A number of genes are needed in the formation of the complicated apparatus involved in hearing, yet any one of these genes which fails to produce a vital part can result in deafness. For instance, one gene may prevent the proper transmission of impulses over the nerve from the ears to the brain, and this gene will be epistatic to all other genes for normal hearing. In other persons the nerve may be all right, but another gene may cause a hardening of the ear bones which must be flexible if they are to properly transmit vibrations to the inner ear. The end result of deafness is the same in both cases, but for two very different reasons. When two persons with hereditary deafness marry, they may have children with normal hearing, if the deafness is caused by different recessive genes. In other families the deafness of the couple may be due to the same gene, and all their children will be deaf.

Thus, we see that human characteristics may be the result of genes having different reactions with one another. In Chapter 9 we will consider the method of gene functioning and get some understanding as to why genes vary as to dominance and recessiveness.

Chapter 5

BOY OR GIRL?

When a woman first begins to experience the symptoms which indicate the development of a new life within her body, one of the first questions to enter her mind is concerned with the sex of the child. "Which will it be—boy or girl?" Such curiosity is understandable because so much of life is affected by sex.

The differences between the sexes is much greater than is commonly realized. Everyone is familiar with differences associated with reproduction and nursing, but they go far beyond this. Skin, muscles, bone, and blood all show differences according to sex. How are genes related to these differences, and what determines which sex a child shall be? These are questions to be answered in this chapter.

THE SEX DETERMINING TRIGGER

The method of sex determination becomes somewhat simplified if we start with an understanding of a basic fact. All persons have all of the potentialities of both sexes. You may be a real he-man with bulging muscles and a hairy chest, yet you have all of the genes needed to produce female characteristics which

53

would make men whistle as you pass. Likewise, the most feminine of girls has genes which, if expressed, would enable her to compete on equal terms with men in any athletic contest. The determination of sex, therefore, hinges on some sort of mechanism which stimulates one set of genes into activity and suppresses the genes for the opposite sex.

A study of the chromosomes of males and females reveals the nature of the triggering mechanism. When cells from a female are spread on a microscope slide and the chromosomes analyzed, it will be found that they can be assorted into twenty-three equal pairs—each chromosome of one size and shape has a counterpart of the same size and shape. Cells from a male, however, show a slight difference—there are only twenty-two matched pairs. The twenty-third pair consists of one long chromosome and one which is very short. This difference accounts for the determination of sex.

The longer of the two chromosomes in the male is the **X-chromosome,** and the shorter is the **Y-chromosome.** Female cells have two X-chromosomes, but no Y-chromosome. These are known as **sex-chromosomes.** The rest of the chromosomes are known as **autosomes.** The reproductive cells have only half as many chromosomes as the **somatic cells** (body cells), so each egg will carry one X and twenty-two autosomes. Sperms, on the other hand, will be of two different kinds. About half of them will carry the X, and the other half will carry the Y. Both kinds will have twenty-two autosomes.

Sex determination now becomes considerably simplified; it depends upon which of the two kinds of sperms fertilizes the egg. If a sperm carrying an X penetrates the egg, the resulting zygote will have two X's, and the resulting embryo will express its genes for femaleness, while the genes for maleness will be held in abeyance. A sperm carrying a Y, on the other hand, will result in the XY combination, and a boy will develop.

HORMONES AND SEX

The chromosome method of determining sex seems very definite, yet we all know that hormones have a great influence on the expression of genes related to sex characteristics. We can in-

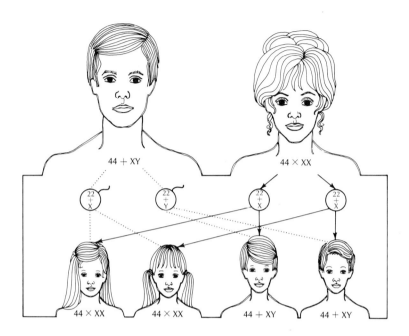

FIGURE 5–1

Sex determination. Two X-chromosomes cause a person to be a female, while an X and a Y result in a male. Sperms may carry either one of the sex-chromosomes, so the sex of children is determined by which type of sperm fertilizes the egg.

ject male hormones into young female chicks, and the latent genes for maleness will be stimulated. The chicks will develop the comb and spurs of the male and will even attempt to mate with other female chicks. Some men, dissatisfied with their sex, have had operations to remove the source of the male hormone and, by taking female hormones, have achieved a degree of sex reversal. Some women with advanced cancer have received treatment with male hormones; this suppresses their own female hormone which seems to stimulate cancer growth. A side effect, however, is beard growth, a change of voice, and expression of other male characteristics as the latent genes for maleness are activated by the male hormones.

How can we square these known facts with the equally well established role of the chromosomes in determining sex? A study of embryonic development gives us a clue. As the embryo is formed, its sex seems to be indeterminate. Reproductive glands appear, but these are the same regardless of the sex into which the embryo will grow. These glands have some cells which can produce testes and other cells which can produce ovaries. Reproductive organs are formed, but these are of such a nature that they can form the organs of both sexes. Then, at about seven weeks of age, something happens that causes the organs of one sex to continue development, while those of the opposite sex remain rudimentary. These rudimentary organs may remain throughout life, but they do not function. The chromosome combination seems to be the activating agent. The XY combination is the trigger which causes the testicular portion of the glands to continue growth, while the ovarian portion stops. Male hormones from the growing testes stimulate the genes for male traits and inhibit those for femaleness. The opposite is true when the XX combination is present.

Thus, by the time a child is born its sex is well established, although it must wait until adolescence before there is a full expression of the genes for a mature male or female. At this time a surge of activity in the hormone-producing cells of the testes or ovaries stimulates the activity of the genes for final maturation of the reproductive organs. Variations in the degree of hormone secretions can result in variations in the extent of expression of the characteristics associated with sex.

WHEN SEX CHROMOSOMES GO WRONG

Unfortunately, the sex chromosomes are not always assorted in the proper manner during the formation of reproductive cells, and abnormalities of sex can result. In rare cases the two paired sex chromosomes may stick together so tightly during meiosis that they fail to separate as they should. As a result, both of them go to one cell and none to the other. If this happens during oogenesis, some eggs are produced with two X's, and some with none. Both types may be fertilized and develop into embryos.

When the two-X egg is fertilized by a Y-sperm, the zygote will have the XXY combination. This has the XX combination of the female, but also has the Y of the male. What sex will it be? It is an abnormal male with a condition known as **Klinefelter's syndrome.** The sex organs are male, but only about half normal size. The muscles are somewhat feminine in appearance, and there may be some degree of breast development. He will always be sterile, and an examination of the testes shows an absence of the cells needed to produce sperms. The extra X has upset the delicate balance of genes which trigger sex, so some of the genes for femaleness are expressed to a certain degree.

It has become customary to make a chromosome analysis of all those entered in women's competitions in the Olympic games and other international track meets. This action was initiated after the discovery that a famous Polish woman athlete had a Y-chromosome as well as two X's. She had the breast development and body build characteristic of a woman, but was actually a male with Klinefelter's syndrome.

When the two-X egg is fertilized by an X-sperm, a three-X zygote results, and this becomes a **super female.** Such persons are relatively normal females, and many females have this three-X condition without ever knowing it. A study of the chromosomes of women in mental institutions, however, shows a higher than expected proportion with this abnormal sex chromosome combination. The chance of lowered mental capacity seems to increase with any of the abnormalities associated with the sex chromosomes.

When the no-X egg is fertilized with an X-sperm, a zygote results with only one X, a total of forty-five chromosomes. This will produce a child with **Turner's syndrome.** Such a child is female, but does not mature sexually at adolescence; the sex organs remain in the infantile state. Also, she will be smaller than normal physically and will have a characteristic large fold of skin along either side of the neck. Many of the children with Turner's syndrome are also mentally retarded. While a single X seems to trigger genes to produce a female body, there are insufficient female hormones to bring about maturation.

When a no-X egg is fertilized by a Y-sperm, the embryo starts development and then dies. Some of the genes on the X-

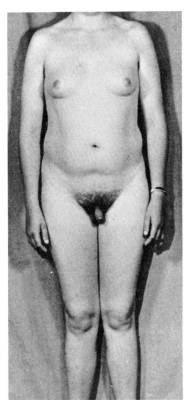 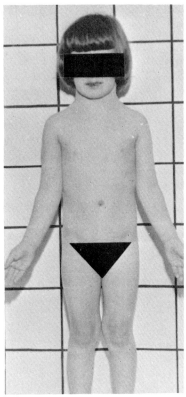

FIGURE 5–2

Non-disjunction of the sex-chromosomes during meiosis can result in abnormalities of sex. The XXY combination results in a person with Klinefelter's syndrome. The sex organs are male, but some feminine traits are expressed. A person with Turner's syndrome receives only one X and is female. The sex organs will not mature, and she will be somewhat smaller than normal. (Left photograph courtesy of Polv Riis and right photograph courtesy of Henry K. Silver)

chromosome are necessary for survival, and no person can live
without at least one X.

The sex-chromosomes may also fail to disjoin in spermatogenesis and result in some sperms with both X and Y and an
equal number with neither. An XY-sperm fertilizing a normal
X-egg results in the XXY and Klinefelter's syndrome. A normal
egg fertilized with the sperm with neither X nor Y will give a
zygote with a single X and Turner's syndrome results.

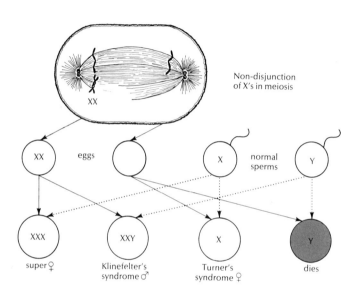

FIGURE 5–3

*Abnormalities of sex which can result from non-
disjunction of the X-chromosome during meiosis of
oogenesis (egg formation).*

Sex-chromosomes And Crime.   The people of the world were
shocked recently when a man named Richard Speck entered the
apartment of nine nurses in Chicago and murdered eight of
them after some had been sexually assalted. In a routine check
of his chromosomes he was found to be XYY. A study of a number of men who had been convicted of similar sex crimes and

murders showed that several of them were also XYY. All of the men were somewhat taller than average and tended to be on the dull side mentally. This led to the suspicion that this particular sex-chromosome combination might result in pathological psychoses which might lead to such aberrant behavior. In fact, in one court case in Los Angeles a lawyer used this argument in defending such a man claiming that it was the extra Y-chromosome which led to excessive male aggressiveness and resulted in the crimes. His client, therefore, should not be held accountable. Fortunately, for our entire judicial system, the court refused to accept this plea to free the accused of responsibility. Had this been allowed, pleas of bad heredity would flood the courts in criminal cases all over the country.

Routine examination of the chromosomes of men leading normal lives shows some cases of the XYY combination among them. Hence, we know that this combination does not necessarily lead to a deviant personality with associated sexually-oriented crimes. The chance for such a personality, however, seems to be much greater when the XYY is present.

How is it possible for a person to get the XYY combination? The various combinations of sex chromosomes which result from non-disjunction during the first division of meiosis do not include XYY. We should remember, however, that there are two divisions to meiosis, and a failure of the sex chromosomes to disjoin in the second division of spermatogenesis can give a sperm with YY. In the first division one cell is formed with an X and another with a Y. If we follow the cell with the Y, we find that it is double and normally this separates to form two Y's, and one goes to each sperm formed. In those cases of non-disjunction, however, both of the Y's go to one sperm and none to the other. When the two-Y sperm fertilize a normal X-egg, the XYY results.

Determining Sex Of Individual Cells.   A few years ago, Murray Barr at the University of Western Ontario was studying nerve cells of cats in an effort to determine if there was any physical change of such cells when fatigued. He found no difference in the cells from fatigued and rested cats, but he did find something else which was very interesting. The cells from female cats had a dark body lying against the nuclear membrane, but no such body was present in cells from males. Soon this investigation was spread to human beings, and it was found that many human cells also showed this difference according to sex. A few cells scraped from inside the mouth of a female will show this dark body, but none are found in such male cells. It became known as the **Barr body** and has important human applications.

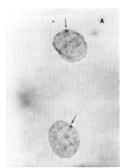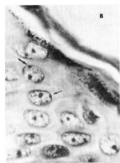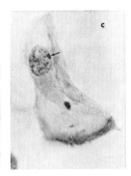

FIGURE 5–4

Barr bodies. These female cells show the darkly stained, tightly coiled X-chromosome lying against the nuclear membrane. Some cells shown are not at the right focus to show the bodies. These cells are: (A) Epithelial cells from the lining of the mouth, (B) Cells in a section of skin, (C) A cell from the lining of the vagina. (Photograph courtesy of Murray Barr)

Why is the Barr body present in female cells and not male cells? It appears that one X-chromosome in each cell uncoils and functions. In the male there is only one X, and this one uncoils. In a female, however, there are two X's, and one uncoils and functions, while the other remains rather tightly coiled and forms the Barr body. The genes in this chromosome do not function because they cannot open up and give out their messages when the chromosome is thus coiled.

This discovery made available a simple method of detecting many abnormalities of sex which were due to irregular distribution of the sex chromosomes. A person with Turner's syndrome, for instance, will be female, but will not have a Barr body; her one X uncoils. One with Klinefelter's syndrome will have a Barr body even though he is a male; one of his two X's will remain coiled. A super female will have two Barr bodies; two of her three X's will be coiled.

Today at many large hospitals a routine check for Barr bodies is made on every newborn baby. This is easily done from the membranes of the embryo, which are generally discarded anyway. The placenta with its attached membranes is obtained. A portion of one of these membranes, the **chorion,** is then spread on a microscope slide, stained, and the cells analyzed for Barr bodies. The advantage of such a check lies in the early recognition of sex abnormalities. Parents of a child with Turner's syndrome may not know that anything is wrong until time for adolescense when they note no normal sexual maturity. By detecting the single X at birth, however, hormone treatment can be given which will make the child more like a normal female. Klinefelter's syndrome can be recognized by the Barr body in cells from a baby listed as male. Unfortunately, it is not possible to recognize the XYY combination by this method; there will be no Barr bodies as is true of the normal male. A complete analysis of all the chromosomes must be made for detection of this chromosome abnormality.

Sex Mosaics And Intersexes.   At carnival side shows one may occasionally see what is purported to be half man and half woman. The arms on one side of the body may be muscular and hairy while on the other they are more feminine. Even the face will be made up to accentuate the illusion. Such mosaics with a bilateral division of sexual traits does occur among insects and some other lower animals, but the sex hormones achieve an even distribution of the sex characteristics in the vertebrate animals including man. Hence, the side show freaks are an illusion built up from men who might be underdeveloped muscularly on one side of the body.

It is true, however, that abnormal distribution of the sex chromosomes during the cell divisions of the early embryo can produce a person with different sex chromosomes in different parts of the body. For instance, a person may show the normal female Barr body in cells from the mouth, yet express all the symptoms of Turner's syndrome. A study of the chromosomes from the region of the reproductive glands, however, may show only one X. Sometimes in mitosis, one X may lag behind and fail to get included in the nucleus which forms, and one of the two daughter cells will have only one X while the other has two. If the cells descending from the one with only one X form the reproductive glands, then Turner's syndrome will result, even though many of the body cells are XX.

In very rare cases the sex chromosome trigger fails to go off on schedule. In an XX person the trigger may be delayed, and there will be some development of the testicular tissue in the early embryo. Then, belatedly, the trigger stimulates the ovarian portion. A person can result with both testes and ovaries, and the effect of both hormones results in an **intersex (hermaphrodite).** Today, surgery is usually performed on such persons early in life so as to remove one of the two types of glands and with hormone treatment a person with fairly normal sex characteristics will result. They will be sterile, however, because of the early effect of the hormones of the opposite sex.

Partial Expression Of Sex Abnormalities.   Sometimes physicians will find a child who shows some symptoms of Turner's syndrome, but they are not as definite as in most cases. Chromosome analysis shows the reason. The child may have one complete X and a second X which has lost a portion. The size of the portion which has been broken off and lost determines the degree of expression of the syndrome. Likewise, degrees of Klinefelter's syndrome result when the person has a normal X and Y plus a portion of another X.

THE WEAKER SEX

Many a young person approaching the age for marriage has wondered if there are enough members of the opposite sex to go around. Such persons can be reassured by a reference to statistics which show that the sexes in the United States are about equal in the age group ranging from about twenty to thirty years. However, the records of births show that about one hundred and six boys are born to every one hundred girls. How can this be when the two kinds of sperms are produced in equal quantities? Are the female embryos weaker than the male embryos, so that more female embryos die before birth? We know that probably at least a fifth of all embryos that start life never make it to a live birth; maybe there are more girl embryos in this group. We can check this hypothesis by examining the great collection of embryos at the Carnegie Institution of Washington. Among these we actually find a slight excess of boy embryos. Hence, we must conclude that more boy embryos start life, and this must mean that more Y-sperm than X-sperm fertilize the eggs. To explain this we might guess that the Y-sperm has some advantage in the journey to, and penetration of, the egg. The Y-sperm might be slightly lighter in weight than the X-sperm. Most of the sperm's head is filled with chromosomes, and the Y-chromosome is considerably smaller than the X-chromosome. Perhaps this slight difference in weight gives the Y-sperm some advantage in the journey up the oviduct to the egg, or perhaps it has some slight advantage in slipping through the cells surrounding the egg.

Girls might think that this excess of live boy babies over girl babies would give them an advantage—they should have a larger number of potential mates from which to choose. But we have already pointed out that the sexes are about equal at the time of life when it is most important that they be equal. This is explained by the fact that boys die at a faster rate than girls. A boy may be stronger physically, but when it comes to survival, it is the girls who are the stronger sex. This greater chance for survival continues to manifest itself throughout life. At fifty years of age there are only about eighty-five men to each one hundred women; at eighty-five the women outnumber the men almost two to one, and at one hundred there are five times as many women as men.

Why should the males be the weaker sex in terms of survival? For one thing, you remember that females have more genes. They have many genes on their extra X-chromosome which males do not have. Of course, the males have a Y-chromosome which the females do not have, but the Y is small and has very few genes, so this is no advantage. If a man happens to carry a harmful recessive gene on his single X, he will express this gene. A woman, on the other hand, may carry the same gene on one X, but she will probably have a dominant normal allele on the other X and will not show the harmful trait. We shall learn more about this in the next chapter. Also, the sexual differences themselves may account for the differences in survival. A male has a higher rate of body metabolism than a female—he uses food and releases energy at a faster rate. Perhaps this causes him to "burn out" more rapidly than a female. Also, the male aggressiveness causes him to take more chances and accidental deaths may be higher.

THE SEX RATIO IN RADIATION STUDIES

Many studies on experimental organisms have shown without question that high-energy radiation, such as is given off from X-ray tubes and from radioactive substances, can cause gene mutations. Most mutations have a harmful effect when expressed. All organisms which can continue to exist on the earth must have established a very efficient set of genes. Mutations are random changes in genes, and anytime you make a random change in something which is already very good, you are most likely to make it less efficient. It has seemed logical that man's genes would also be subject to mutation as a result of exposure to high-energy radiation. His genes have the same type of construction as those of all other forms of life, and there is no reason why he would be different in his response. Still, it would be desirable to have direct evidence. He is such a slow-breeding creature and has so few children, relatively speaking, and we cannot experiment with him as we can other animals. Also, since most mutations are recessive to their normal alleles, it takes several generations to detect most mutations which have been induced.

Faced with such obstacles, geneticists came up with a plausible possibility. Males have only one set of genes on the X-chromosome, and they will express any lethal genes on that X. Mutations include many lethal genes, so anything which in-

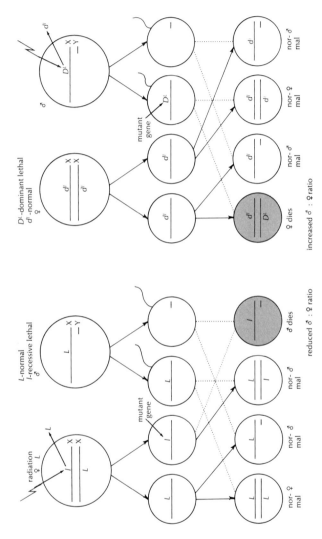

FIGURE 5-5

How radiation can affect the sex ratio. When women receive the radiation, they have fewer sons because some male embryos may die from induced recessive lethals. Men who receive the radiation tend to have fewer daughters because some female embryos may die from dominant lethals. Most lethals are recessive, so the effect is more pronounced when women are radiated.

creases the mutation rate would be expected to reduce the proportion of males to females in the first generation. If a recessive lethal mutation is induced on an X of a woman and she passes that X to a son, he will die. If a daughter receives it, she will live because she will get an X with the dominant normal allele from her father. Hence, if many women receive heavy radiation, and it causes some lethal sex-linked recessive mutations, you could detect a smaller than average number of males in their offspring. What about dominant lethals? They are less common than recessive lethals, but they would result in a decrease in girls when the fathers receive the radiation because girls receive an X from their father, but boys do not. Dominant lethals induced in the mother would kill boys and girls both.

Turpin in France made a study of the percentage of males born to over four thousand patients who had received extensive X-rays in the course of gastrointestinal examinations and tumor treatment. They had received from 70 to 270 roentgens which is over a hundred times what a person receives when he has a simple chest X-ray picture made. The results were:

Women radiated: males in offspring before radiation, 54.1%
                    males in offspring after radiation,   48.5%
Men radiated: males in offspring before radiation, 51.5%
                    males in offspring after radiation,   56.1%

The decrease in the male ratio after radiation of women and the increase of the ratio after radiation of men are indicative of the induction of lethal mutations on the X-chromosome.

A similar study was made of the children of the survivors of the atom bomb tragedy in Hiroshima, Japan. The average amount of the radiation received by these people was much less than in the French studies, but there was a slight indication of the same type of result in the first studies.

Chapter 6

HEREDITY INFLUENCED BY SEX

There is an old belief that girls are more like their fathers and boys like their mothers with respect to inherited characteristics. As is true of many such beliefs, there is a little truth involved, but mostly it is false. Girls receive an equal number of genes from both parents, so this part of the belief is entirely unfounded. Boys inherit equally all of the genes on their forty-four auto-somes (all except sex-chromosomes), but there is a difference with respect to the sex chromosomes. They receive their single X from their mother and will express most of the genes on this chromosome. Hence, with respect to this fraction of their total genes, they do inherit more from their mother. Of course, boys receive a Y-chromosome from their father, but this is such a small chromosome and carries so few genes that it cannot compare with the X in the total picture of heredity.

Genes on the X-chromosome have an interesting pattern of inheritance. A man may show a trait because he carries such a gene, but it does not appear in any of his children. Then, perhaps to his surprise, the trait appears in a son of one of his daughters. It will never reappear in the sons of his sons. This is sometimes called a "skip generation" method of inheritance.

SEX-LINKED (X-LINKED) INHERITANCE

Most of the genes on the X-chromosome do not have alleles on the Y-chromosome. These are known as sex-linked genes, and the traits they produce are called sex-linked traits because the pattern of inheritance is related to sex. They may also be called X-linked genes and traits because the genes are on the X-chromosome. Males will express all the sex-linked genes they carry, even though some are recessive. A recessive is not expressed when it has a dominant allele. Since males have only one set of sex-linked genes, they express them all, dominant and recessive alike.

Colorblindness.   A boy once went in a candy store and, pointing to a candy jar, asked for a green sucker. The clerk took out a green one, but the boy said, "No, not the red one, I want that green one." He was pointing to a red sucker. He was discovering, as many boys do each year, that he was colorblind, that he could not readily distinguish red from green. People have long observed that this trait is much more common in boys than in girls and when the principle of sex-linked inheritance was discovered, the reason became known. One of the genes on the X-chromosome is involved in the production of small bodies, cones, in the retina of the eye. These cones are sensitive to green light and send a signal to the brain whenever green light strikes them. A recessive allele of this gene produces defective cones which cannot detect green clearly, so green objects are often confused with red objects.

Figure 6–1 shows how the recessive gene for this type of colorblindness is inherited. A boy can show the trait only if his mother is a carrier of the gene. All of a colorblind man's daughters will be carriers, but will not show the trait unless they also get the gene from their mother.

Human geneticists in pursuing investigations of colorblindness have found that it is somewhat more complex than was at first supposed. The type we are describing is known as the green insensitive or **deutan** type. There are two varieties of the gene which deviate from normal. One causes the extreme form of

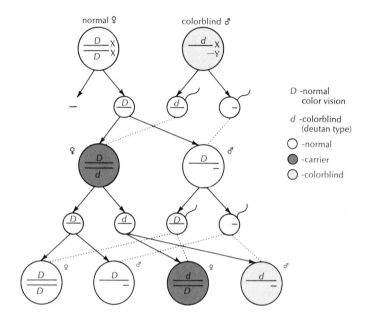

FIGURE 6–1

Sex-linked inheritance as illustrated by colorblindness of the deutan (green insensitive) type. Males have only one of each of their sex-linked genes and therefore express them all, even though some are recessive.

color defect which we have described and is known as **deuteranopia.** This gene renders the cones practically insensitive to green. Another allele, however, causes a milder form of color vision defect. It seems to produce cones which are defective, but not completely insensitive to green. Boys expressing this gene can see green if it is a bright color and the light is good, but have trouble with pastel shades and when the light is poor. They are said to have **deuteranomaly.**

About 75 percent of the persons with red-green colorblindness have one of the two deutan types. The other 25 percent have the red insensitive type, or **protan** colorblindness. This also

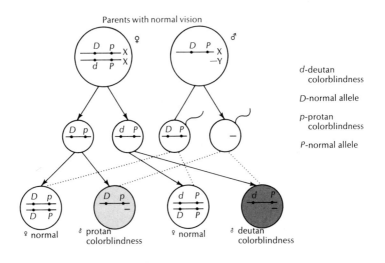

FIGURE 6–2

Normal parents can have sons with both types of colorblindness when the mother carries a gene for each type on her X-chromosomes.

exists in two types, the extreme **protanopia** and the milder **protanomaly.** This is due to variations in a gene at another location on the X-chromosome. Colorblind boys usually never know which of the two types they have; both cause a confusion of red and green, and it is only through special tests that the distinction can be made.

Boys have a much greater chance of being colorblind than girls. A boy will show the trait if he receives one of the genes on his single X-chromosome, which has come from his mother. A girl will be colorblind only if she receives the same gene from both parents, which means that her father must be colorblind and her mother must carry the same kind of gene. A girl can even have normal vision if both father and mother are colorblind. The father may have the protan type and the mother have the deutan type. As a result, she will receive only one gene for each type from her parents, but she will also receive a dominant normal gene for each type.

Statistics show that about 8 percent of the boys in the United States have one of the types of red-green colorblindness, but only about 0.4 percent of the girls will have such defective color vision. This is about one boy in twelve, but only one girl in two hundred and fifty. Total colorblindness has also been found, where the cones lack sensitivity to all colors, but this is quite rare.

The Bleeder's Disease. A few years ago the crown prince of Spain, Prince Alfonso, was driving down one of the busy streets of Miami, Florida. Another car pulled in front of him, and there was a screeching of brakes and a sudden crash as the cars collided. It was not a serious accident, as automobile accidents go, but the prince was cut slightly by the flying glass. The cuts were bleeding profusely, however, and he died from loss of blood before he reached a hospital.

This death came about because the prince had the inherited disease of **hemophilia,** the "bleeder's disease." It is a disease which was well known to the royal families of Europe, for there had been ten cases of boys who had it in the past three generations. It started from a recessive sex-linked gene carried by Queen Victoria of England and was spread to other countries of Europe by marriage of her children into other royal families. It even got into the Russian royal family, and the little Tzarevitch Alexis had many a close brush with death from bleeding before he and the other members of the royal family were killed by the communists. In fact, the Russian revolution itself may have been precipitated by this one gene. An unscrupulous monk, Rasputin, seemed to be able to stop the bleeding when Alexis was at the point of death, and as a result he was given great power in Russia. Many feel that his abuse of this power and his oppression of the Russian people were important factors in the precipitation of the revolution which finally ended with the communists in power.

Persons with hemophilia are deficient in a part of the blood plasma which is needed for normal clotting. Normal blood usually clots within five minutes after a wound causes it to flow from the blood vessels. The clot seals the wound and stops the bleeding. In persons with hemophilia, however, the clotting may be delayed as long as two hours. Such persons must always be very careful because even a small bruise may cause internal bleeding which continues for hours.

Hemophilia is extremely rare in females. Only about one male in ten thousand is born with hemophilia. There is an equal chance that a female will receive the gene on one of her two X's, and the same chance that she will receive it on her other X. Hence, about one female in five thousand will be a heterozygous carrier of the gene. Thus, there is very little likelihood that a hemophiliac man will marry a carrier woman. Mathematically, the chance would seem to be 1/10,000 X 1/5,000 or one in fifty million. Half of the daughters of such a marriage would have hemophilia, and this is only one in one hundred million. Even this figure is too frequent, however, because about three-fourths of the males with hemophilia never live to marry and have children. On the other hand, the chance increases when there is close inbreeding in a family which carries the gene. In England three cases of hemophilia were reported from a region where there were many marriages of close relatives. These cases seemed to be caused by a form of the gene which creates a delayed clotting of the blood, but not the extreme delay found in some cases. Were it not for this, the girls probably could not have survived the monthly menstrual cycle.

The method of inheriting hemophilia has been known since ancient times. In the Hebrew families of Biblical times there were reports of boys who died from bleeding when the operation of circumcision was performed. This was considered a religious rite and was performed on all boy babies, but special exceptions were made for boys born to women who were in a family where such deaths had occurred. No such exception was made where the father was from such a family. They had learned the basic method of inheritance of this trait.

As is true of colorblindness, prolonged bleeding can result from several causes, and several different genes may be involved. The form which we have described is the so-called classical hemophilia, **hemophilia A,** resulting from a defect of a sex-linked gene which has the task of producing one of the plasma proteins, the antihemophilic globulin, needed in the blood clotting reaction. There is another gene at a different location on the X-chromosome which causes excessive bleeding for another reason. This gene results in a deficiency of another plasma component, the plasma thromboplastic component. The result is **hemophilia B** or **Christmas disease,** and is milder in its effects than hemophilia A. Then, there is **pseudohemophilia** resulting from a dominant gene on an autosome. This gene interferes with the breakage of the blood platelets which contain a substance needed in clotting. Still another gene, which is autosomal recessive, interferes with the production of fibrinogen in the plasma, and fibrinogen is the material which is converted into the fibrin which makes the clot. This condition is known as **afibrinogenemia.** Thus, we see that sometimes a number of different genes can have the same end result, although for different reasons.

Sex-Linked Traits Expressed Only In Males. Some recessive genes on the X-chromosome have such an extreme effect when expressed that they cannot be found in females. These cause the death of the males who receive the gene before they can reproduce. Such a gene causes one of the most common types of muscular dystrophy, **pseudohypertrophic muscular dystrophy.** A boy with this gene has a normal early childhood, but at about ten or eleven years of age the muscles begin to waste away until the boy seems to be literally skin and bones. Death comes by the mid teens. No woman carrier could ever bear a child from a man with the trait, so a girl could never receive such a gene from both parents, barring the most unlikely event of a mutation in a man's reproductive cell. This has never been observed.

Sex-Linked Traits More Common In Females. Many people think of sex-linked traits as being peculiar to males, but we have seen that females can also show them and, in fact, there are some that are found more often in females. When a gene on the X-chromosome is dominant, it will be expressed twice as often in females as in males. This is because a female has twice as many chances of receiving this gene as a male; she may receive it on the X from her father or on the X from her mother. A male, on the other hand, has only one X from his mother, so he has no chance of receiving the gene from his father. **Defective**

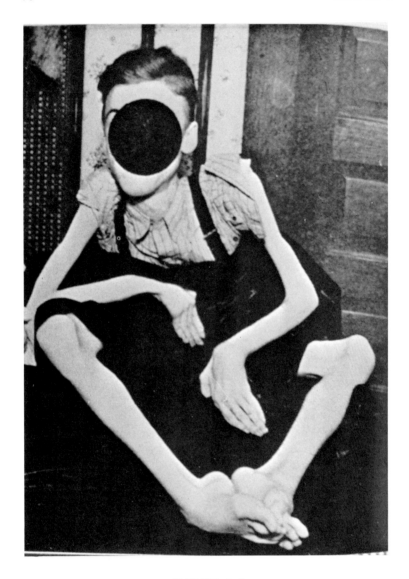

FIGURE 6–3

Pseudohypertrophic muscular dystrophy, a sex-linked trait. This teen-age boy received the gene from his mother. It causes, first a swelling, and then a wasting away of the muscles about the time of adolescence.

dentine is a trait caused by such a dominant gene. Anyone with this gene wears down the teeth rapidly and usually has only stumps of teeth protruding from the gums. A man with this trait can pass it on to his daughters, but not to his sons.

Women As Mosaics Of Sex-Linked Traits. A woman is a mosaic of sex-linked traits. Some cells of her body may be expressing one sex-linked trait, while other cells are expressing an alternate trait. This is due to the coiled state of one of her X-chromosomes in each cell of her body. As we learned in Chapter 5, one of the two X's in each cell does not open out and function; instead it remains tightly coiled and forms the Barr body. Hence, each of a woman's cells has only one functioning X, as it true of the cells of a man. Then, you may wonder, why do not some women heterozygous for a sex-linked recessive gene express the trait. The answer is that some of her cells are expressing the dominant allele and some the recessive allele. Take hemophilia; for instance, a woman heterozygous for the gene will have some cells which can make the important plasma component needed for clotting. In these cells the X with the normal gene is uncoiled and functioning. Other cells have the other X functioning, and these cells cannot produce the component. Those that do, however, can produce enough of it to bring about normal clotting. These cells just have to work twice as hard. Delicate tests on women who have had hemophiliac sons, however, have shown that these heterozygous women do have a slightly extended blood clotting time. A few cases have been found where such women had a considerable prolongation of blood clotting time. By chance, a large proportion of their functining X's carry the gene for hemophilia.

The same is true of colorblindness. Some of the cells of the retina of a woman heterozygous for the deutan type can perceive green, while other cells cannot. Since the cells are so small, however, the woman thinks that she sees green with all her cells. Delicate tests of her retina, however, may reveal certain areas which are insensitive to green.

This principle of inactivation of one X was worked out in mice by Mary Lyon to explain the mottled black and yellow condition found in some females heterozygous for the sex-linked genes for black and yellow. It is now known as the **Mary Lyon hypothesis.**

Incompletely Sex-Linked Genes. The Y-chromosome of the male does have a few genes which are alleles of some genes at one end of the X-chromosome. Without some such matching of genes, the two chromosomes would not pair in meiosis. Such genes will be present in pairs in both males and females and are called **incompletely sex-linked genes.** A very small proportion

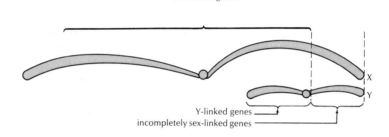

FIGURE 6–4

Relationships of genes on the X-chromosome and the Y-chromosome. Most genes on the X have no alleles on the Y; these are sex-linked genes. A few genes are allelic on both chromosomes; these are the incompletely sex-linked genes. A few are found only on the non-homologous portion of the Y; these are the Y-linked genes.

of the total genes on the X are thus involved, however, and a very few such genes have been found. One such gene causes a type of blindness because a pigment forms over the retina and another causes a type of kidney disease.

Y-Linked Genes. A few genes are also on the Y-chromosome which have no alleles on the X. These include the genes which

play a part in triggering the development of male characteristics and perhaps a few more. One study of hairy ears in India
seems to show that it is always passed from a father to all his
sons, and this is exactly what would be expected of a gene on
this portion of the Y-chromosome.

SEX-LIMITED GENES

Sex-limited genes are those which are normally expressed in only
one sex. They may be on any of the chromosomes and should
not be confused with sex-linked genes. The beard is a good example. A woman does not normally have a beard, yet she carries all of the genes necessary to produce a beard. Also the type
of beard which her sons develop depends on the genes he receives from his mother as well as from his father. A man may
have a soft silky type of beard, yet his son may have a tough,
wiry beard because of a dominant gene inherited through his
mother. On rare occasions some of these genes may show in the
wrong sex. The bearded lady at the carnival may be a normal
woman in other respects, but this one trait, which is normally
limited to males, is expressed. Likewise, the development of the
breasts is normally limited to women, but there are rare cases
where there is some breast development in otherwise normal
men. The sex hormones seem to influence the development of
sex-limited characteristics in most instances at least.

SEX-INFLUENCED GENES

As you sit in the balcony of a theater and look down on the audi-
ence below, you may be impressed with the rather large number of bald heads which are so prominent among the male
members. Disease may cause a loss of hair, but the great majority of people who are bald are that way because of their
genes. It is quite evident that this is a trait which is much more
common in men than women, although it is known in women.
Hence, it must be related to sex, although it is neither sex-linked

nor sex-limited. Baldness seems to result from a gene which is dominant in men and recessive in women. A man expresses the gene when he has only one of them, but a woman must have two. A woman has another advantage—she can hide her baldness much easier than can a man. With all the wigs which women with perfectly normal heads of hair are wearing, it is a simple matter for a woman who is a little short of nature's head covering to substitute an artificial thatch with no one the wiser.

Another trait which seems to be the result of a sex-influenced gene is the length of the index finger (the one with which you point) as compared with the fourth finger (the ring finger). In

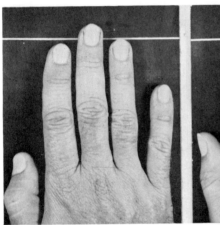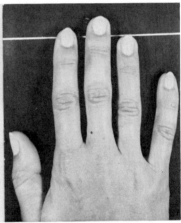

FIGURE 6–5

A sex-influenced trait. An index finger which is equal to, or longer than, the fourth finger is a dominant trait in females, but is recessive in males. Both hands shown here are of persons heterozygous for the gene. The female hand, right, has the long index finger, but the male hand, left, does not.

some people the index finger is shorter than the fourth finger, and in some it is equal to or longer than the fourth finger. If you study the fingers of a rather large group of people, you will nearly always find more boys than girls with the short index finger, although it is found in both sexes. If we assume that the trait is due to a gene which acts as a dominant in males and a recessive in females, we can explain the difference in frequency in the two sexes. As an example of how the children would come out in a single family, assume that a woman has the short finger and her husband has the long index finger. All of the sons of this couple would have the short finger, and all the girls would have the long index finger. This is because the woman would be homozygous for the gene for short index finger, and her husband would have to be homozygous for the gene for long index finger. All of the children would be heterozygous, and the girls would show the long finger, but none of the boys would show it.

Thus we have seen that the sex of a person can have an important bearing on the expression of many genes, some of which have nothing to do with the sexual characteristics.

Chapter 7

BLOOD WILL TELL

"Blood will tell" is an old saying which implies that a person's heredity will come out in the end despite any attempts to modify the basic inherited tendencies. The idea originated from an ancient belief that blood was the material of heredity and that a new life originated from a blending of blood from the two parents.

As geneticists unraveled the secrets of inheritance, this old concept was disproved; now we know that blood has nothing to do with the transmission of hereditary traits. A new meaning has been uncovered, however, for the old saying, "blood will tell." Even though all blood appears to be very much alike upon superficial examination, we have found that it shows a great many inherited variations. These are so numerous that your blood is almost as distinctive as your fingerprints. If a small sample of your blood were to be thoroughly analyzed today and then, years from now, a sample of your blood along with samples of a thousand other persons were to be analyzed, it would be possible to pick your sample from the entire group. Others in the group would share many of your blood traits, but no other one would have exactly the same combination of these traits that you have. A wounded burglar who left behind a few drops of blood, could later be identified by the particular combination of blood characteristics which he possessed. Thus, blood *will* tell when it is used as a means of identification. Also, blood will tell when there is a dispute as to paternity of a child. An examination of the blood of the mother, the child, and any men who might be the father will show which one has actually contributed genes which have established the blood characteristics of the child. Blood will also tell when there has been a mix-up of babies in a hospital. Analysis of the blood of the parents and the babies involved will show definitely to whom each baby belongs.

Blood is a favorite of human geneticists because the method of inheritance of its many recognizable differences is so clearcut and easily recognized. Many of man's external physical features, such as the shape of the nose for instance, show so many quantitative variations and may be subject to such environmental modification, that it is often difficult to establish a gene for trait relationship.

THE BLOOD ANTIGENS

Blood **antigens** are among the major inherited blood characteristics. What is an antigen? It is the part of a protein molecule which can stimulate **antibody** production. We can explain with a simple demonstration using a guinea pig. Inject the guinea pig with some of the raw albumen (white) of a chicken egg. The guinea pig shows no reaction to this injection. Wait ten days, however, and give it a second injection of the same thing, and it will probably go into spasms and die. What happened during those ten days to cause this reaction? The albumen of the egg is a protein and, therefore, contains an antigen. This stimulated the guinea pig to react to the first injection by producing antibodies which were highly specific for chicken albumen. Upon the second injection the antigens reacted with the antibodies so violently that death resulted from spasms of the heart muscles. We can watch the reaction under the microscope by mixing some of the blood serum from the guinea pig with some egg albumen. There will be a precipitation of the albumen, a cloudiness forms, because of the reaction.

Antibody production is a protective mechanism. When foreign proteins in the nature of disease germs enter your body, you produce antibodies to destroy them and prevent future infections. Vaccinations bring about immunity by introducing weakened or dead germs into your body. When a person needs a kidney transplant or becomes sensitive to penicillin or pollens in the air, he may wish that he did not produce antibodies, but survival would be difficult without this ability.

DISCOVERY OF BLOOD TYPES

If you are an average person weighing about one hundred and fifty pounds, you have about five quarts of blood in your body. It is such a vital fluid that you will die if you lose as much as about three pints of this fluid at one time. Man has long known of the importance of blood to life, and there were crude attempts to save lives by blood transfusions as early as the eighteenth century. Some of these efforts were successful—some men with dueling wounds, which ordinarily would have been fatal, were saved by passing blood directly into their veins through small tubes leading from the veins of other men. In other cases, however, when the transfusions were made in the same way, the persons receiving the blood died as soon as the bloods began to mix.

Later it was found that when blood from two different persons was mixed in a container outside the body, it would mix smoothly in some cases, but in others the blood cells would stick together in clumps and separate from the clear plasma. In an effort to explain this difference a man named Landsteiner removed the cells from the plasma and recombined the cells and plasma from different people. In some cases a smooth combination resulted, but in others the red cells clumped. He found that these reactions were due to antigens in the blood cells and that there were two antigens, which we now designate as A and B. Many persons have neither antigen and are known as type O. Some have cells with only the A antigen and are known as type A. Some have only the B antigen and are known as type B. A few persons have both antigens and are known as type AB. Every person also has all the antibodies in the plasma

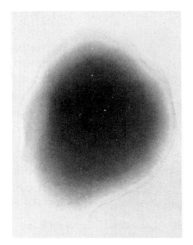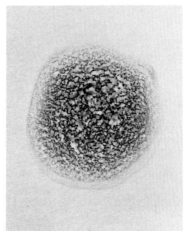

FIGURE 7–1

Unclumped and clumped red blood cells. Both samples are from a person with type A blood. The sample at left has been mixed with serum containing b antibodies; there is no clumping. At right, some of the blood has been mixed with serum containing a antibodies; the red blood cells have agglutinated into large clumps.

which he can have without clumping his own blood. Thus, type O persons have antibodies *a* and *b* which will clump antigens A and B respectively; type A persons have *b* antibodies, and type B persons have *a* antibodies, while type AB persons do not have either kind.

Before a person is given a blood transfusion, his blood type is carefully checked and suitable donor blood is chosen so there will be no clumping of cells. In most cases a person is given blood of the same type as his own, but in an emergency it is possible to give certain other types as shown in Table 7–1.

You will note from this table that type O blood can be given to a person with type A, but A cannot be given to O. This is possible because as a transfusion is given, the antibodies coming in with the plasma are rather quickly diluted by the plasma already in the body and will not have a high enough concentra-

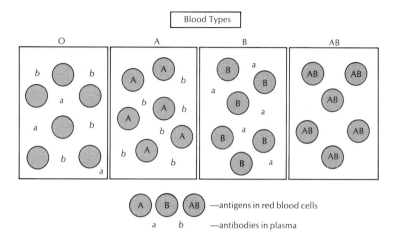

FIGURE 7–2

Antigen and antibody characteristics of the four types of blood. Each type has all the antibodies possible without clumping its own red blood cells.

TABLE 7–1

Blood Transfusion Donors and Recipients

Blood Type	Can Donate to:	Can Receive From:
O	O, A, B, AB	O
A	A, AB	O, A
B	B, AB	O, B
AB	AB	O, A, B, AB

tion to cause serious clumping. Hence, the antibodies a in type O blood will not cause serious complications when given to a type A person. The reverse transfusion, however, will be fatal because the A antigens in the red blood cells being given to a type O person will be clumped by the high concentration of a antibodies in the O blood.

METHOD OF INHERITANCE OF THE ABO BLOOD TYPES

What determines a person's blood type? A set of three allelic genes are responsible for the four basic blood types. One of these, designated as *A* produces the A antigen as a part of the protein in the plasma membrane around the red blood cells. An allele of this gene, designated as A^B, produces the B antigen. A third allele, designated as *a*, produces neither of the two antigens. The first two alleles described are co-dominant, so a person will be type AB if he receives both of them. The third allele is recessive to the other two. Table 7–2 below shows the gene combination which results in each of the four blood types.

TABLE 7–2

Types of Genes	Blood Types Resulting
a/a	O
A/A or *A/a*	A
A^B/A^B or A^B/a	B
A/A^B	AB

This shows how the antigens are determined, but what about antibodies? We know that every person will have all the blood type antibodies which he can have without clumping his own cells. Where do these come from? In all other known cases the plasma will not have antibodies that react with a foreign antigen unless the person has contacted that antigen. In this case, however, the antibodies will be present without any apparent contact with the antigen. A type A person will have *b* antibodies even though he may never have seen a type B person. The human fetus and newborn infant do not produce antibodies, so no matter what the blood type there will be neither a nor *b* antibodies in their plasma. At about two to eight months of age, however, these antibodies begin to appear, although the child will be about eight years old before they reach their maximum level (titer). What stimulated their production? One hypothesis holds that they are simply inherited like other traits with a delayed expression. Since no other antibodies are known to be inherited, however, this is open to question. A more

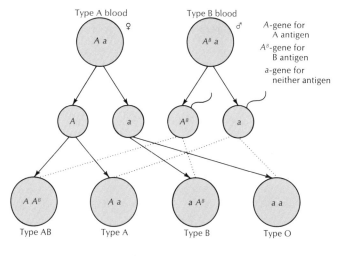

FIGURE 7–3

Blood types to be expected in the children of a type A woman married to a type B man, when both carry the recessive gene for O.

plausible hypothesis holds that the embryo contacts many antigens upon his entry into this world, and some of these may be very similar to the A and B antigens. Since a person never makes antibodies which react with his own antigens, he will not react to those very similar to A if he is type A. If he is type B, however, he will react and produce antibodies to A-like antigens in his food and other contacts. These will react with the A antigens in blood cells of type A persons and are known as the a antibodies.

SOURCE OF THE A AND B ANTIGENS

The great majority of people have at least one of a dominant gene, *H*. This produces substance H which may be transformed into either A or B antigen. If you have the gene, *A*, you produce A antigen from substance H, while the gene A^B produces the B antigen from this substance. You cannot produce either antigen unless you also produce the precursor, H. In Bombay, India, a few persons were found who could not produce H; they were homozygous for the recessive allele, h. A study of the pedigrees

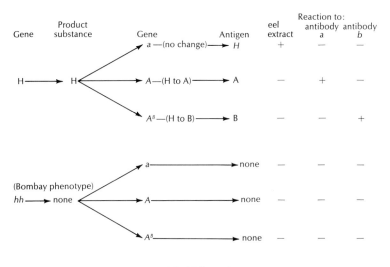

FIGURE 7–4

The relationship between the H substance and the A and B antigens. Those with the recessive genotype, h/h, have the Bombay phenotype. They cannot pro-duce substance H, so have nothing to convert into A or B antigens even though they may carry genes for A or B.

of these people showed that some had to carry the genes for A or B antigens, yet their red blood cells would not react with a or b antibodies. They were said to have the **Bombay blood type.**

Both the Bombay and the O blood types have cells which are not clumped by a or b antibodies. How can we distinguish between the two? In the course of trying many different extracts to test their effect on human blood, it was found that a serum of common eel blood would clump the cells of a type O person. Likewise, an extract of a plant, *Ulex,* will do the same thing. Actually, what these extracts react with is the substance H. A person with the Bombay type would not show any reaction to these extracts because he has no substance H.

A valuable offshoot of this discovery has been the ability to recognize heterozygous type A and B persons. A person who is A/A will use most of the substance H to produce A antigen, and his cells will not react to the extracts. One who is A/a, however, will not use all the substance H, and the cells will be clumped by the extracts. The same is true of B type persons. This can be used in some cases of disputed parentage. Suppose a type O woman bears a type O child, and a type A man is sued for support of the child. He could be the father if he carries the gene for O, but not if he is homozygous A/A. If his red cells clump when exposed to the extract, he does carry the gene for O and could be the father. No clumping would rule him out.

THE SECRETOR TRAIT

You may read in some "cloak and dagger" mystery stories of a detective identifying a man from the dried saliva on an envelope he had licked to seal. This may seem to be one of the far-fetched cases found only in fiction, but it is actually possible to make a degree of identification from such a body secretion. About 77 percent of the American people are **secretors,** they possess a dominant gene, Se, which causes their blood antigens and the H substance to enter many of the body secretions, such as saliva, gastric juice, tears, and milk (if a nursing female is involved). Those homozygous for the recessive se, do not have such antigens.

How can we recognize the antigens in the body secretions? There are no cells to clump as there are in the blood. Let us assume that a person is a type A secretor. When we mix a antibodies with his saliva, there is a reaction. We cannot see it, but the antigens and antibodies have neutralized each other. If we now add some type A cells to this mixture, there will be no clumping. When a person is not a secretor, the saliva-antibody mixture will clump type A cells because the antibody is still active. The H substance will also be in the secretions and can be demonstrated by adding the extracts of the eel or the plant and then cells from a type O person.

Thus, you can tell a lot about a person from his saliva: whether or not he is a secretor, and, if he is a secretor, what type of blood he has.

THE RH FACTOR

After the human blood types were discovered, people began testing the blood of other animals to determine if they also had blood types. One such animal was the Rhesus monkey. Red blood cells from one of these monkeys were injected into some guinea pigs. After about two weeks blood serum was obtained from the guinea pig blood. When red blood cells from a Rhesus monkey were mixed with this serum, the cells clumped. The guinea pig had produced antibodies against some antigen which is found in the cells of Rhesus monkeys. This antigen was called the **Rh factor.**

Next the investigators wondered if man shared the gene for this factor with the monkeys, so human blood was mixed with the guinea pig serum. The cells from some people would clump, but others showed no reaction. Hence, some people have this gene for the Rh factor and others do not. Those that do are called **Rh-positive**, and those who do not are called **Rh-negative.** The Rh factor was found to result from a dominant gene, and its recessive allele causes people to be Rh-negative.

This discovery led to an understanding of a problem of human transfusion. For many years physicians had been plagued with unexplained reactions. A man with type O blood needed a transfusion. A type O donor was found, and a successful transfusion was given. Later, perhaps this same man needed a second transfusion, and the same donor was used, but this time the red blood cells being transfused were clumped with perhaps fatal results. The reason was clear when it was found that the recipient was Rh-negative and the donor was Rh-positive. The first transfusion was successful because people do not have natural *rh* antibodies, but the introduction of the positive cells sensitized the man so that he had *rh* antibodies when the second transfusion was given. Today no negative person is given a positive transfusion, even though it is the first.

The Rh Factor And Childbirth. Another medical mystery was solved by the discovery of the Rh factor. For many years doctors had found that a certain percentage of the babies which they delivered had a severe anemia, and their red blood cells were deformed and sometimes nucleated. They had **hemolytic disease,** or **erythroblastosis.** Sometimes these babies died within

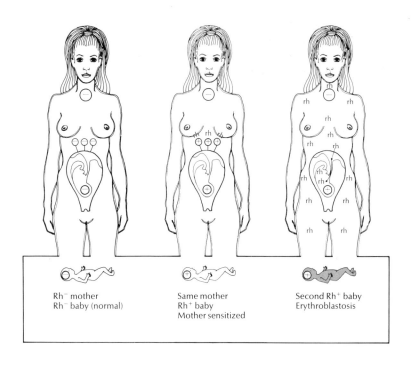

Rh⁻ mother
Rh⁻ baby (normal)

Same mother
Rh⁺ baby
Mother sensitized

Second Rh⁺ baby
Erythroblastosis

FIGURE 7–5

Erythroblastosis, or hemolytic disease of infants, can result when an Rh-negative woman is sensitized by bearing an Rh-positive child at a previous birth.

a few days of birth, but if they could survive for as long as a week, they gradually replaced their defective cells with normal cells. Strange to say, these babies almost never appeared at a first birth, but became more numerous with each additional birth. Once having borne one such baby a woman had about a fifty-fifty chance of having additional babies of this nature. Heredity and various environmental factors were postulated as possible causes. An understanding of the Rh factor provided the correct answer. These defective babies were born to Rh-negative women with Rh-positive husbands. It was evident that the blood of such a woman's first positive child was sensitizing her, and in future positive children, some of her *rh* antibodies entered the fetal blood stream and reacted with the fetal blood cells with the resulting abnormality.

Doctors learned to predict the chance of this hemolytic disease and gave the baby blood transfusions at birth to help

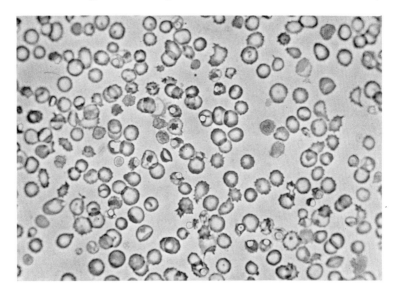

FIGURE 7–6

Blood from a baby with erythroblastosis. Note that many of the red blood cells have nuclei. They have been released prematurely and are poor transporters of oxygen.

overcome the anemia. Sometimes transfusions were given to the embryo before birth. When stethoscope examinations showed that the baby was having difficulty and might not survive until birth, the doctor might inject normal blood directly into the fetus through the abdomen of the mother. In extreme emergency the abdomen might be opened, and the baby taken out, transfused with blood, then replaced in the uterus to complete its development.

The question arises of how can a positive baby sensitize its negative mother if the placenta provides a barrier which prevents an actual mixing of the blood cells. It is true that cells do not pass through whole placental membranes, but during the latter period of pregnancy there may be small ruptures of these membranes. A few of the red blood cells of the fetus can thus cross the barrier. Then, at birth, there is massive bleeding of both maternal and fetal tissue as the placenta breaks loose, and quite a number of fetal cells can gain entrance into the mother's blood. A fetus has a different kind of hemoglobin from an adult, and these cells can be identified in a sample of the mother's blood after a birth. These cells may sensitize the mother. Some negative women have been known to give birth to several positive children without any difficulty. Whether or not she is sensitized depends upon the number of fetal cells she receives and her ease of sensitization. Another factor plays a part. This is the ABO blood type of the mother and her child. When a mother has antibodies which react with the cells of the fetus, these cells may be destroyed before they can sensitize the mother. Let us say a type O, Rh-negative woman has a type A, Rh-positive child. When the type A positive cells enter the mother, her a antibodies destroy the cells before the Rh antigen can sensitize her. If both mother and child are type O, then she has a much greater chance of being sensitized. Statistics bear out these findings.

Today there is no reason for any woman to become sensitized, nor should there be any more Rh-induced hemolytic disease. We have developed a preventive called **Rhogam** which can be injected in a negative woman who has just borne her first positive child. This Rhogam consists of rh antibodies which react with and neutralize the Rh antigens in fetal cells which have gained entrance to her body. Hence, she will not produce any *rh* antibodies. For women already sensitized it has no value.

Rhogam can be obtained from Rh-negative men or post-menopause women who volunteer to be sensitized by receiving carefully controlled injections of the Rh antigen. Blood serum from these persons will then be rich in *rh* antibodies. These antibodies are concentrated in the protein gamma globulin fraction of the blood serum. This fraction is extracted, and it is the Rhogam.

Complexities Of The Rh Factor. Continued study of the Rh factor showed that it exists in three different forms, which are actually produced by three different genes which lie very close to one another on the chromosome. These genes are designated *CDE,* and their recessive alleles are *cde.* A person is Rh-negative only if he is homozygous for all three recessives, *cde/cde.* If any one of the genes is a dominant, the person will be Rh-positive. They could be *cDe/cde,* for instance. The D antigen is the one of primary concern, because about 95 percent of the positive people in America include this antigen in their make-up. It also happens to be more highly antigenic than the others and more readily causes sensitization. C is next in frequency and antigenic strength, while E is the least frequent and least antigenic. Some positive people have two of the antigens, say C and D, and some carry all three.

In blood transfusions the check for the Rh factor is often only for antigen D, using blood serum of a laboratory animal sensitized to the blood cells from a person who had only the D antigen. Those positive for C or E would not be detected by this method, but they are treated as if they were negative and are given negative blood. A few cases are known where hemolytic diseases developed from C sensitization of a woman who was positive for D alone and bore a child positive for C or CD.

To complicate things still further the recessive genes *c, d,* and *e* seem to result in some weak antigens which can some-times cause a low degree of antibody production. A person who is *CDe/CDe,* for instance, might have some difficulty with a second transfusion of blood that is *cDe/cDe.* This is possibly a situation similar to the H substance for the ABO blood types. Each of the three Rh antigens may have a precursor substance, and each of these can stimulate antibody production in some laboratory animals. Hence, we can get anti-c, anti-d, and anti-e, which can be used to tell whether a positive person is heterozygous or homozygous for all three genes. Credence to this postulation was furnished by the recent discovery of a few persons who were Rh-null. They might inherit genes for C, D, or E, but show no positive antigens, nor will their blood react with the antigens against c, d, or e. They must express a gene or genes which fail to produce a precursor for the Rh antigens.

OTHER RED BLOOD CELL ANTIGENS

The ABO and the Rh antigens are by no means all of the antigens which can be demonstrated on the surface of human red blood cells. An **M and N series** was found by injecting human red blood cells into guinea pigs. Two basic antigens were found resulting from two co-dominant alleles. Thus, people can be divided into three MN blood types, M, N, and MN. Since people do not produce antibodies against these antigens, they have no clinical significance. Later it was found that both the genes for M and N had slightly variant alleles, so the number of possible combinations was increased.

The **Xg antigen** was discovered when a Mr. And received a transfusion of blood and began showing some reaction even though all known antigens matched. It was found that he was Xg-negative while some previous donors of blood had been Xg-positive, and he had become sensitized to the antigen, Xg. The gene for the antigen is dominant and located on the X-chromosome.

A Mrs. Kidd bore a child with mild symptoms of hemolytic disease, yet there was no Rh incompatibility. She turned out to be homozygous for a recessive gene while her husband carried the dominant allele for an antigen which is now known as the **Kidd-factor.** The **Diego antigen** is interesting because it is found in about 10 percent of the Japanese people, but practically all Caucasians are Diego-negative. Other antigens which have been identified are: Auberger, Dombrock, Duffy, I, Kell-Cellano, Lewis, and Lutheran. Only rarely do these have medical significance.

OTHER BLOOD VARIANTS

Hemoglobin Variants. All of the blood variants considered so far in this chapter have been of the antigens on the surface of the plasma membrane of red blood cells. The number increases greatly when we go to other parts of the blood. For instance, inside the red blood cells is the hemoglobin, a red compound which is necessary for the transportation of oxygen from the lungs to the cells of the body. This is a protein with a molecule consisting of an iron compound in the center from which extend four long chains of amino acids. Many varieties of these amino acids are known, all resulting from the action of various genes. These are discussed in Chapter 10.

Leukocytes. Leukocytes (white blood cells) do not show many inherited variations when compared with the erythrocytes, red blood cells. They do show an interesting distinction according to sex. Granular leukocytes, those with granules in the cytoplasm, typically have a nucleus formed of two or three large lobes. In females, in addition, there is a **drumstick-shaped projection** from one of the lobes. This seems to represent the coiled X-chromosome and is the equivalent of the Barr body in other female cells. **Pelger anomaly** is a recessive trait characterized by an abnormal shape of the leukocyte nuclei. One form of **leukemia,** an excess of leukocytes, has a genetic basis and is discussed in Chapter 10.

Thrombocytes (Blood Platelets). These are very small blood cells, usually only about half the size of red blood cells. They normally burst when they contact a rough surface as would occur when a wound is inflicted. They contain a chemical substance which initiates the clotting reaction. Before they can burst, however, they must adhere to a solid surface. One autosomal recessive gene, when homozygous, prevents the normal adhesiveness of the cells so they do not burst readily, and clotting is retarded. The trait is known as **pseudohemophilia** or **Von Willebrand's disease.**

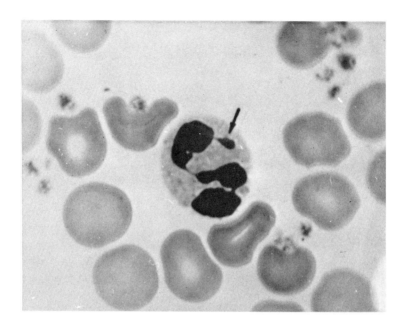

FIGURE 7–7

A drumstick extending out from the nucleus of a leukocyte from a female. Male leukocytes do not have such drumsticks.

The Plasma Proteins.   Now let us turn our attention to the liquid part of the blood, the plasma. We have already learned how heredity can cause variations in some of the plasma components related to blood clotting (see Chapter 5). Most of the inherited plasma differences, however, are to be found in the plasma proteins. Three of these proteins are known as **globulins;** there are alpha, beta, and gamma globulins. Also, there is **fibrinogen,** which is involved in blood clotting, and **albumin,** the most abundant of the blood proteins. Variations in these proteins are recognized by starch-gel electrophoresis. A bit of the protein is placed on the jelly-like starch-gel, and it is placed in a buffered solution and subjected to a strong electrical field. The different

proteins and their variants migrate for different distances because of the variations in the size of the molecules and their
electrical charges.

   Haptoglobins are **alpha globulins** that bind free hemoglobin
which is released when red blood cells burst. It is transported to
the liver where it is removed and broken down. This is an important function because free hemoglobin in the blood is toxic.
Most people have one of three haptoglobin types resulting from
two co-dominant alleles. The alleles, Hp^1 and Hp^2, give the three
types, $Hp^{1\text{-}1}$, $Hp^{1\text{-}2}$, and $Hp^{2\text{-}2}$. Three degrees of variation have
been found for each allele, so there can be as many as eighteen
different combinations. A different allele, Hp^0, has been found
which is recessive to the other two and produces no haptoglobin at all. Those homozygous for this gene are constantly
suffering from the toxic effects of free hemoglobin in the blood.

   Transferrins are **beta globulins** which transport ionic iron
from the intestine, where it is absorbed, to the body parts where
it is needed. Free iron ions in the blood are toxic, yet the body
needs for iron are great, and this solves the problem of transportation. Transferrin C is the most common and results from
the gene, Tf^c, but fourteen different alleles have been found.
All seem to function equally well, so the primary value of this
knowledge is in establishing patterns of migration, intermar-
riage, etc. There is an interesting variation in different races.
About 8 percent of the Navajo Indians have $B_{0\text{-}1}$ transferrin, yet
this is practically unknown in Caucasians. About 10 percent of
the Americans with a Central African ancestry have the D_0 type,
but only a fraction of 1 percent of those of European ancestry
have this type. Was it chance or some factor in the environment
which caused these differences? These are important questions
for population geneticists.

   Immunoglobulins, gamma globulins, contain the antibodies. Two gene loci have been identified which are involved in
producing these. One, the *Gm* locus, has been found to have a
variety of alleles which have different electrophoretic patterns.
The second, *Inv* locus, also has a number of such variations.
These variations are not only of value in population studies, but
the ability to react to foreign antigens is related to the gamma
globulin types. Thus, heredity can account for a person's ability
to resist infection from disease germs as well as his tendency to

develop allergies. Also, a recessive gene is known which causes **disgammaglobulinemia,** a condition in which gamma globulin is produced, but it cannot react with foreign antibodies. Hence, the homozygous person is an easy prey to infections, but he has no troubles with allergies. Another recessive gene causes **agammaglobulinemia** which does not produce any gamma globulin at all.

Fibrinogen is the plasma protein which solidifies into fibrin in the clotting process. If one recessive gene fails to produce this protein, the homozygous person has **afibrogenemia.** Clotting cannot take place, so bleeding from injuries is severe. It does stop eventually however as the wells of the injured vessels adhere.

With only this incomplete list of blood variations it is easily understood why the blood is a very distinctive characteristic and why an exact duplication of any one combination of blood traits could rarely occur.

Chapter 8

THE NATURE OF GENES
AND HOW THEY CHANGE

Did you ever wonder why you do not have eyes on the bottom of your feet or ears growing out of your shoulders? You have genes for eyes and ears in all of the cells of the body, yet normally these organs develop only at the particular place they are supposed to develop and nowhere else. We have much to learn about how genes are controlled so that they work where and when they should, but we have learned something about the process. To understand this, we must first learn more about the nature of the genes and how they direct the activities in the cells in such a way that particular characteristics are expressed. These are some of the things to be considered in this chapter, along with the important topic of gene mutation, or how genes change and then reproduce themselves in the changed condition.

THE LADDER OF LIFE—DNA

In 1962, the Nobel prize was awarded to two men for their discoveries on the nature of the gene. James Watson and Francis Crick, using a special type of X-ray pictures along with chemical and physical analysis, concluded that the gene was somewhat like a long, flexible ladder which was twisted around itself many times. Each twist of the ladder included four rungs. This ladder was found to be made of **deoxyribonucleic acid.** Since it is such a long name it is usually referred to as **DNA.**

The flexible supports of the DNA ladder are made of molecules of a simple sugar **(deoxyribose sugar)** held together by bonds of phosphate. The rungs on the ladder are made of four types of compounds or bases. These bases are joined together in pairs, two to each rung. Two of these are always found together; they are known as **adenine** and **thymine,** which we shall abbreviate to **A** and **T.** The other two are also always found together; they are **cytosine** and **guanine,** abbreviated to **C** and **G.**

If genes are thus made of such a few parts, you may be wondering how it is possible for there to be so many different kinds of genes. One human being will have many thousands of genes in each cell, and while there will be a few duplicates in this group when a person is homozygous for particular genes, most of these genes will be different from one another. Then, we must remember that there are many other kinds of genes in other persons. Also, man is only one of about two million different species of living things found on the earth. The number of different kinds of genes in all forms of life, all the way from viruses and bacteria up to man, must be so great as to be almost beyond comprehension. How can this be possible when the number of parts used to make genes is so limited?

Such a variety of genes is possible because of the great length of the DNA ladder which forms a single gene. There may be as many as one thousand to thirty thousand rungs on the ladder which makes one gene, and a change in just one rung can make a gene operate in a different manner. If one rung, say **AT,** is taken out and replaced in a reverse position, **TA,** the entire gene is altered. Also, a substitution of **CG** for **AT** would alter the gene. Such changes sometimes do happen; they are known as **mutations.** Thus, we can see that there are four possible types of rungs on the ladder: **AT, TA, CG,** and **GC.** From these four we can make an almost limitless variety of combinations, considering the length of the ladder. We might compare it to making words with up to thirty thousand letters using a four-letter alphabet. Some have compared it to the Morse Code which had only two symbols, dots, and dashes, yet it can form any word in any language. Even a very long word will have no more than thirty or forty dots or dashes at the most.

THE DUPLICATION OF THE GENES

We learned in Chapter 2 that genes duplicate by becoming unjoined in the middle of the strand, somewhat like a zipper becoming unzipped. Then, each half gene attracts to itself the parts needed to form the other half of the "zipper." Within the cell are free molecues of **A, T, C,** and **G,** as well as the sugar and phosphate needed to construct the missing parts. These have come from digested food, acted upon by cellular enzymes. Each **A** on a half gene attracts to itself a **T** along with a molecule of the sugar and phosphate. Likewise, each **T** attracts an **A, C** attracts **G,** and **G** attracts **C.** Thus, two new genes are formed, each an exact duplicate of the original. This is shown diagramatically in Figure 8–1.

HOW GENES WORK

Most of the metabolic activities of a cell take place in the cytoplasm, and these activities are directed by the genes which lie within the nucleus. How do the genes impart the information needed by the organelles in the cytoplasm? They do so by messages which pass from the nucleus to the cytoplasm. In the cytoplasm are small bodies, **ribosomes,** which receive the gene messages, **messenger RNA,** and assemble amino acids into long chains, known as **polypeptide chains.** Such chains may be a protein, or several such chains may be combined to form a protein. This is all any gene can do, direct the formation of proteins; all of the results of gene action can be traced to this simple activity.

The proteins produced are of two basic types. **Structural proteins** are those which become a part of the protoplasm of the cell and are, therefore, necessary for cell growth. **Functional proteins** are in the nature of enzymes which promote chemical

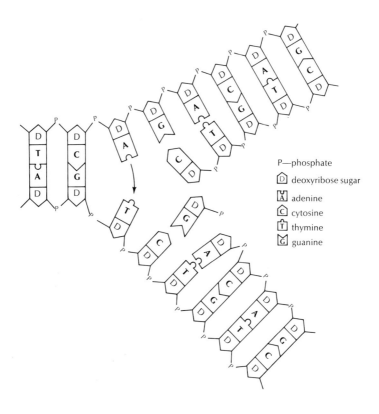

P—phosphate
Ⓓ deoxyribose sugar
Ⓐ adenine
Ⓒ cytosine
Ⓣ thymine
Ⓖ guanine

FIGURE 8–1

Diagram showing the parts which make up a gene and
how these parts are put together. It also shows how a
gene duplicates by splitting where the bases join and
then how each half gene attracts to itself the missing
portions.

changes within the cell. Most of the proteins produced are of
the latter type. Each cell must be able to produce thousands of
enzymes in order to carry on its complicated reactions.

We can now understand why some genes are dominant,
others recessive, and still others co-dominant or intermediate.
A **dominant gene** is usually one that directs the production of a

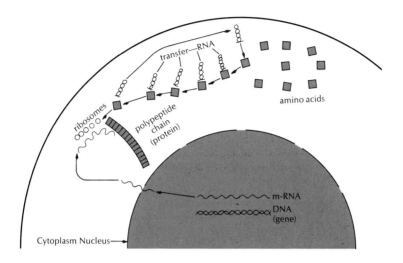

FIGURE 8–2

*How messages from the genes pass out of the nucleus
to the ribosomes and code the formation of a polypeptide chain by assembling amino acids.*

specific protein, either structural or functional. A **recessive
allele** of such a gene either does not produce the protein or
produces it in a defective form. The gene for normal pigmentation, *A*, for instance, produces the protein enzyme which converts a precursor substance into melanin. The recessive allele, *a*,
does not produce this enzyme. In a heterozygous person the one
dominant gene can make all the enzyme needed, but in a person
homozygous for the recessive allele, the enzyme is not produced, and the person is an albino. In a few cases, a dominant
gene may result in the absence of a cell product because the
dominant gene produces something which inhibits the formation of the product. There is a rare form of albinism which is
due to a dominant gene which produces something that inhibits cell absorption of one of the products needed to form
melanin.

In the case of the A and B blood antigens, both the gene
for A and the gene for B produce their antigens when a person is
heterozygous. Hence, the genes are **co-dominant.**

In the case of **polygenic inheritance,** we often have a number of genes, each of which may contribute to the expression of the trait. Let us assume that there are eight genes which can influence the intensity of pigment in the skin. A very dark-skinned person might have all eight genes, four pairs of alleles, each of which makes the pigment deposit a little more intense. A very fair-skinned person may have all eight genes for low pigment deposit. We would then find nine different degrees of degree of skin pigmentation. Studies made in some of the Caribbean Islands of the children of parents with mixed racial ancestry seem to show segregation into nine different shades.

THE GENETIC CODE

Much genetic research has gone into the job of trying to crack the genetic code with a high degree of success. We now know the code for each of the twenty amino acids which are put together to form proteins. Three bases of the DNA chain form the code word, **codon,** for one amino acid. Each messenger-RNA chain moving from the genes will carry a transcription of this code out to the ribosomes. If the code is for a protein made of one polypeptide chain containing one thousand amino acids, then the messenger-RNA will have three thousand bases, three for each amino acid. A change in any one letter of this entire three thousand will change one codon into a different three-letter combination, and one amino acid out of the one thousand might be different. Such alterations do occur and are known as **mutations.** Even though one amino acid out of one thousand may not seem like a very significant change, still this can have far-reaching effects on a person who produces this altered protein. If the protein is an enzyme, the altered form may not be able to carry out its task, and the cells may be lacking some vital product. This could have a lethal effect if the product is essential to life. If the altered product is a structural protein, the effect on the cell and the entire person can also be very great.

The code for the human **hemoglobin** molecule is one of the best understood. This molecule consists of an iron compound as a base to which are attached four polypeptide chains.

They are in pairs, each pair having 287 amino acids. One of each pair is known as the **alpha chain,** coded by one gene, while the other is the **beta chain,** coded by another gene. By digesting the molecule and then using electrophoresis, it has been possible to separate the amino acids and to determine the sequence of these amino acids on each chain. Since the codon for each amino acid has been worked out in lower organisms, this makes it possible to know the sequence of the base pairs in the two genes. It also makes possible the detection of mutations, which result when one of the amino acids is replaced by another. This would happen when one base pair out of the total of 861 in the two genes is altered. For instance, one pair which reads **TA** might be turned around to read **AT,** and the triplet codon would then be different. Or, **CG** might be substituted for **TA.** In one particular case, it is known that the middle pair of a triplet which codes the amino acid glutamic acid is turned around, and the new triplet codes the amino acid valine. Thus, valine is substituted for glutamic acid at a particular point on the beta chain of the hemoglobin molecule. Even though this is only one out of 287 in a half molecule, the results are far-reaching. The hemoglobin tends to form in chains causing the red blood cells to become sickle-shaped, and the person expressing this gene has sickle-cell anemia, as was described in Chapter 4.

AGENTS WHICH CAUSE MUTATIONS

All organisms have naturally-occurring mutations. They are infrequent, but they do take place. Certain agents have been found which can increase the rate of gene mutation. The first to be discovered, and still the most important, is high-energy radiation. **Gamma rays** are wavelengths of energy, similar to visible light rays, but of a much shorter wavelength and with much more energy. As a result they can pass through an animal body, creating ionizations within the cells. The resulting chemical changes may cause some genes to make mistakes in transcribing the genetic code, and a mutation has occurred. Radiation of this nature is given off from X-ray tubes and from radioactive substances, such as cobalt-60. Other types of high-energy radiation consist of actual sub-atomic particles given off from radioactive substances. These may be electrons, neutrons, protons, or combinations of these. They also cause ionization within the cells and mutations.

The number of mutations induced by high-energy radiation is in direct proportion to the amount of radiation striking the cells. Hence, a little radiation to many people may result in just as many mutations as a large amount of radiation to a few. This has caused concern over the radioactive particles which are spread all over the earth whenever a nuclear bomb is exploded in the atmosphere. The amount received by any one person is infinitesimal when compared to the natural background radiation he receives anyway, but over three billion people are exposed. We all receive radiation all the time, cosmic radiation from outer space and radiation from radioactive particles in our surroundings. To this must be added the necessary radiation from medical examinations and treatment, but any unnecessary exposure should be avoided if possible.

Quite a number of chemical agents have been found to cause gene mutations when they are applied to lower forms of life. These include: mustard gas, ethyl-urethane, nitrous acid, nitrogen mustard, phenol, formaldehyde, manganous chloride, and bromouracil. Even caffeine and theobromine, which are in coffee, tea, and chocolate, have been found to be mutagenic when given in very large amounts to some lower organisms, such as bacteria and molds. None of these chemicals seem to pose a threat to man. They do not seem to reach a sufficient concentration in his cells to bring about mutations. LSD is one which has come under suspicion, however, because it has been shown to cause chromosome abnormalities in the blood cells of heavy users and, generally, anything that thus affects the chromosomes will also cause gene mutation.

RESULTS OF SOMATIC MUTATIONS

Usually we think of mutations as they take place in reproductive cells, or their precursors, since these are transmitted to future generations. They can also take place in somatic cells, but they are usually overlooked because most of them do not have any phenotypic effect. If you are heterozygous for the recessive gene for albinism and the dominant allele mutates to albinism in a cell of your intestine, you will be homozygous for albinism in this cell. Even though this cell divides and makes a small clone of cells, you will never know it. Should the mutation occur in a skin cell you may get a tiny albino spot, but you probably would never notice it. Somatic mutations which occur in the early embryo, however, can produce comparatively large areas which show the mutant phenotype. Figure 8–3 shows a boy who evidently had such a mutation while he was a small embryo. As a result he has an area around one eye which is albino. It is conceivable that somatic mutations in cells of the adult which influence genes related to growth control could result in cancerous growth. Agents which cause mutations are also generally agents which can cause cancer.

CONTROL OF GENE ACTION

A very important question relating to gene action remains: what turns them on and off? Obviously, all genes cannot function all the time; this would result only in an undifferentiated blob of cells instead of a human being. Probably no more than 5 percent of your genes are now functioning. These are used in the maintenance of the body processes and repair of worn out or injured tissues. Some of your genes are now switched off, but will be turned on again when the need arises. Others are permanently switched off. Those which formed your brain cells, for instance, will never operate again. You have all the brain cells you will ever have, and any loss remains a permanent loss.

Extensive research in lower organisms indicates that certain **operator genes** turn on the structural genes which produce the messenger-RNA which, in turn, stimulates the production of enzymes and structural proteins. A **regulator gene,** however, produces a **repressor** which holds the operator gene in check most of the time. Only when some sort of inducer comes into the cell and blocks the repression is the operator gene able to stimulate the genes into functioning. As an example, let us say a cell has a gene able to produce the enzyme lactase, which can digest milk

FIGURE 8–3

*Result of a somatic mutation. This boy has a small area
of albino tissue which includes his upper eyelid, part
of the eyebrow, and the surrounding skin. He is heterozygous for the gene for albinism. In the early embryo the dominant allele, A, mutated to the recessive
a in one cell. This cell thus became homozygous, a/a,
and the tissue around the eye descended from this
cell.*

sugar, lactose. As long as there is no milk in the diet, there is no
need for such an enzyme, and none is produced because the repressor holds the operator in check. When lactose enters the
cell, however, it acts as an **inducer** to block the repression, and
the genes pour out the messages for the formation of the enzyme

which will act upon the lactose. When the lactose is digested, there is no more inducer, and the genes shut off.

Hormones are powerful inducers which block the repressor and cause the operator to allow the genes to function. A tissue culture can be made of cells from the lining of the human female uterus. When the female hormone, estrogen, is applied to these cells, they show a great spurt of growth. Tests show that there is a great increase in messenger-RNA output. Studies of embryos of many lower animals show that the early embryo produces chemical **organizers** which cause the proper organs to develop at the proper place. At one point on one side of the head the organizers act as inducers which permit functioning of the genes for eyes. Some of this inducer applied to other parts of the embryo can likewise induce an eye. In this manner, a frog can be made with an eye in the center of his back.

Potentialities Of Gene Control. Armed with this knowledge, geneticists are now pressing forward with investigations on human control of gene action. If we can just learn the exact combination of chemicals needed, we might be able to grow human organs outside the body for transplants. For instance, a person may have a badly damaged heart. We can give him a transplant, but healthy, living hearts are hard to come by. Even when one is transplanted, the recipient tends to reject it sooner or later because it has a different protein composition from his own. But, suppose we knew the secret and could take a few cells from his blood or other body part and give them just the right stimulating chemicals in the proper sequence. We might produce a young, healthy heart which could then be transplanted with no fear of rejection because the recipient is actually using his own tissue, with all of his own protein peculiarities. Or, more likely, it might be that we could inject the chemicals into the person's heart and stimulate it to regenerate the damaged parts with no need for a transplant. We have learned how to re-activate cells to produce body parts, such as legs, in salamanders, frogs, and lizards; who knows but what the process may one day be extended to man.

Chapter 9

ENZYMES—THE AGENTS OF
THE GENES

EFFECTS OF ENZYME DEFICIENCIES

A young couple, Mr. and Mrs. Wright, were expecting their first child. When it was born, they were very happy that it was a bright, physically normal boy. The child continued to develop normally for several weeks, then some alarming disorders began to appear. The child became dull with little interest in his surroundings; his stomach became bloated, and he would often vomit up his milk shortly after feeding. The liver became enlarged, and a cloudiness developed in the lenses of the eyes. Mrs. Wright became greatly alarmed because she remembered an aunt who had borne a similar child. That child developed these same symptoms and later became severely mentally retarded; he became blind from cataracts in the lenses of the eyes, and he suffered from frequent digestive upsets. He died at four years of age. However, the Wright's story has a happy ending. Since the birth of the child to the aunt and this birth to the Wright's, the cause of and a means for the prevention of the condition has been discovered.

When Milk Is Poison.   The condition described above is known as **galactosemia** and is caused by a recessive gene. Both of the Wrights carried this gene, and it became homozygous in their

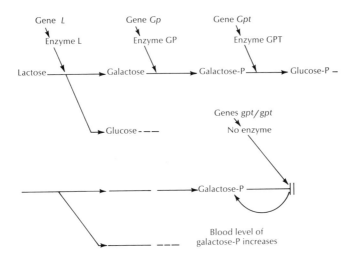

FIGURE 9–1

Galactosemia results when a mutant gene fails to produce an enzyme needed in the series of enzyme-mediated reactions affecting lactose, milk sugar. In this case it is the accumulation of a product which is not broken down that causes serious difficulties.

first child. This gene fails to produce one of the enzymes in the liver which breaks down galactose, a simple type of sugar. Galactose is not generally found in the human diet, so this might seem of little consequence until we learn that milk sugar, lactose, is digested in the intestine into two simpler sugars, glucose and galactose. When this galactose is absorbed and passed through the liver, most of it is broken down into glucose and other compounds by a liver enzyme, **UDPG.** Those who lack this enzyme cannot make the conversion, and the galactose flows through the body in the blood and plays havoc with many body processes. The symptoms of galactosemia are the result of this disruption of the normal processes. A baby who inherited one of these recessive genes from each parent will be normal as long as it remains in its mother's uterus. The mother carries a dominant gene for the production of the enzyme, and this keeps down the quantity of galactose in both mother and child. Once the embryo breaks free from his umbilical bond with the mother, he is on his own. Milk will be the first food he takes, and the galactose will begin to build up. He will seem normal at first, but then the symptoms of galactosemia will begin to appear. Recovery will be complete if the child is taken off milk within a few weeks. Even cataracts in the eyes will clear up. If, through ignorance of the cause of the trouble, the baby is fed on milk for as long as six months, there will be permanently impaired vision and mental defects.

In the case of the Wrights, the baby was immediately taken off milk, both human and cow's milk. Mrs. Wright learned how to prepare a synthetic milk made from an extract of soybeans. The baby's recovery was dramatic. Within a week the abnormality was only like a bad dream.

Normal carriers of the recessive gene for galactosemia can be recognized by a simple test. The person to be tested drinks some water containing galactose. A few hours later the blood is tested for galactose level. It will be higher than normal if he is a carrier. A person with two dominant normal genes can produce enough enzyme to convert this sudden inflow of galactose, but with only one dominant gene he cannot convert it all, and some excess gets into the blood stream. Thus, in those cases where galactosemia is known in a family, a young married couple can know whether or not there is a chance of the enzyme deficiency in their children.

Other such success stories have resulted from the discovery that genes exert their influence on cell activities by means of enzymes which they code. Recessive alleles of any of the genes which code a vital enzyme can result in extreme deviations from normality when these recessive genes are homozygous. As we learn which enzyme is missing and just what that enzyme would do if it were present, we can frequently devise means to prevent the damage. A number of other cases of such prevention will be described in this chapter.

Why Enzyme Deficiencies Cause Abnormalities. An enzyme deficiency can result in a break in a series of chemical changes and cause abnormalities for several reasons. First, as illustrated

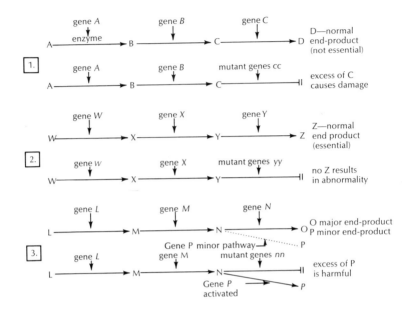

FIGURE 9–2

Three different causes of abnormalities from enzyme deficiencies. (1) Too much of a product which is not broken down, (2) Lack of an essential end product of the enzyme action, and (3) Too much of a substance which is produced in an alternate pathway.

in the case of galactosemia, there may be a build up of an excess of the substance the enzyme is supposed to act upon, and this substance is harmful in such high concentrations. The human blood stream contains many substances which must be maintained within narrow limits of quantity. Too much or too little can cause trouble. A small amount of galactose is actually needed for the brain to develop normally, but this can be obtained from other carbohydrates in the diet. When milk is included in the diet, however, too much galactose is formed from the milk sugar, and this is why we must have the liver UDPG. This enzyme cuts down the galactose concentration in the blood by converting most of it to glucose and other compounds. When this enzyme is not there, the excess galactose pours out into the main blood stream and acts as a poison which prevents the normal functioning of many body cells. Many of man's inherited abnormalities can be traced to such an excess of a compound in the blood because he could not produce an enzyme to break it down. Prevention of this type of abnormality consists of withholding the substance which the body cannot break down.

Cretinism. In other cases, an abnormality develops because the body is deficient in something which the enzyme normally produces. **Cretinism** is a condition which appeared rarely, but regularly in past times. A cretin has a small body, does not develop sexually, has a sloping forehead and remains at the idiot level of mental development regardless of training. In past times, there was nothing that could be done for cretins; most of them lived their lives out in mental institutions. Today we can prevent the development of these symptoms and produce a person of normal stature and mentality. This became possible when we learned the cause. In a normal person the thyroid glands produce the thyroid hormones. Iodine is a basic ingredient of such hormones. Free iodine is harmful in the blood, however, so it is carried combined with an amino acid, tyrosine. At the thyroid glands there is an enzyme, **deiodinase,** which splits the iodine from the tyrosine and releases the iodine needed for the synthesis of thyroid hormones. A recessive allele of the gene fails to produce the enzyme, deiodinase, and fails to release the iodine. Homozygous persons, therefore, cannot produce thyroid hormones, and cretinism develops. It is known as **genetic goitrous cretinism.** Cretinism can develop in other persons if the diet is deficient in iodine during the early stages of childhood development.

Gout. Gout is a condition which, for some reason, is thought to be humorous. A man with his foot wrapped in bandages and propped high in the air may seem funny to others, but to the man involved it is anything but funny. He suffers excruciating

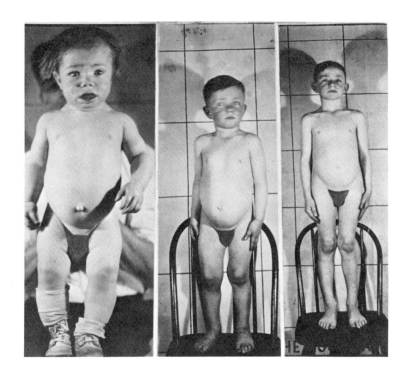

FIGURE 9-3

*Cretinism is an abnormality resulting from a lack of
an end product, thyroxine. The child at left shows
typical symptoms of cretinism. By taking thyroid extract daily, he turned out to be a normal boy, as
shown in the photographs at right. (Photograph
courtesy of Good Samaritan Clinic, Atlanta, Georgia)*

pain with the slightest movement of swollen, irritated joints. The
joints of the big toe are the most frequent targets. Tiny, needle-
like crystals accumulate there and pierce the surrounding tissue whenever there is any movement. This condition represents
a third type of human abnormality resulting from an enzyme
deficiency. When an enzyme is missing to act upon one compound, that compound may be carried along an alternate pathway and result in an excess of an alternate substance.

Amino acids are nitrogen-containing compounds, and
when a person eats more protein food than is necessary to pro-

vide the amino acids needed in cell metabolism, the excess amino acids will be acted upon by liver enzymes. A nitrogen-containing compound will be split off, and the balance of the amino acids can be used for energy needs of the body. The nitrogen compounds are a waste and must be excreted from the body. In man the primary nitrogeneous waste is **urea** which is carried to the kidneys by the blood and excreted in the urine. Enzymes are needed to form this urea. We also have enzymes to convert the wastes into **uric acid,** but this is a minor pathway, and only a little uric acid is produced normally. In a person who has defective genes for the production of urea, however, the enzymes for the alternate pathway are used, and the uric acid level of the blood rises. When it becomes as high as 9 mg per 100 ml, some of it will combine with sodium to form sodium urate crystals. These tiny crystals tend to settle by gravity which typically carries them to the feet including the joints of the big toes. Gout results.

Control of gout can be achieved by restricting the protein intake, so that it does not greatly exceed the amino acid needs of the body. There are also medicines which tend to prevent the crystals from forming. Gout is more common in men than in women. For some reason, men naturally have a little more uric acid production than women, so it is easier for them to reach the critical level. Gout can also result from the action of other genes. One of these, which is a dominant, reduces the ability of the kidney to remove uric acid from the blood.

Some Beans Can Be Deadly.   The inhabitants of Sardinia, an island just off the coast of Italy, have learned to be afraid of fava beans. They can remember times when many of their boys and a few of their girls became very sick after eating fava beans, and there were some deaths. Some even became sick while working in the fields where they inhaled pollen from the fava bean plants. Far away, on an island in the South Pacific, during World War II, a group of newly-arrived recruits were given a new drug, primaquine. This drug had been found to be very effective in preventing malaria, one of the great health hazards in this region. Within a few days, however, a few of the men became highly anemic and had to be shipped back. Most of the men so affected had an ancestry traceable back to Africa, Italy, Greece, or some of the other areas in the Mediterranean region. What possible relationship can these two events have?

As you might expect, the same enzyme deficiency is the explanation for both events. An enzyme, known as **G6PD (glucose-6 phosphate dehydrogenase)** is found in the red blood cells of most persons. It is produced by a dominant gene on the X-chromosome. The enzyme is one of a series needed for obtaining energy from glucose, the most common source of cell energy. A recessive allele fails to produce the enzyme, and the cells generally are normal, but there is some product of fava beans, primaquine, sulfa drugs, and a few other products which acts as a poison to the cells if this energy is not readily available. This is an example of an enzyme deficiency which does harm only in certain situations. This condition is known as **G6PD deficiency, favism,** or **primaquine sensitivity,** depending on the agent causing symptoms. (*Note:* Fava beans may be called broad beans or even lima beans, since they are closely related to lima beans.)

ENZYMES IN SERIES

Most of the enzyme-stimulated changes in cells take place as a series of small changes, each resulting from one enzyme. Hence, any major change usually requires the action of a number of enzymes, each carrying the reaction only a small part of the way. Each enzyme is produced by one gene, and a mutation of any one of these genes can result in a recessive allele which cannot produce a normal enzyme. Thus, the chain of events is broken when a person is homozygous for such a gene. One of the best understood of such a series of changes in man is related to the breakdown of the two amino acids, **phenylalanine** and **tyrosine.** Practically any protein food you eat will contain both of these, and as the protein is digested, it is broken down into its component amino acids which are absorbed in the blood stream. They can then be absorbed into cells where they are exposed to cellular enzymes in a series. First, an enzyme converts the phenylalanine into tyrosine, and then through the action of a number of enzymes the tyrosine can be converted into other compounds along any one of four pathways. This is shown in Figure 9-4. We now know of four breaks which

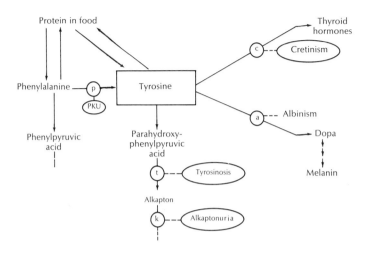

FIGURE 9–4

The series of enzyme-mediated reactions involving the amino acids, phenylalanine and tyrosine. Enzyme deficiencies have been discovered at four points in the chain of reactions.

may occur in the series of changes, each with a different effect according to where the break takes place. Let us consider these.

PKU—Too Much Of A Good Thing.

The first break can come in the conversion of phenylalanine to tyrosine. A recessive gene fails to produce this enzyme, the necessary liver enzyme, and there is a build up of phenylalanine in the blood. Some of it goes along the alternate pathway to phenylpyruvic acid, so this also becomes higher than normal in the blood. The kidneys begin to remove these, and they can be detected in the urine. The afflicted person has **PKU, phenylketonuria.** All persons must have some phenylalanine; it is an amino acid essential to the production of proteins by the cells. Too much though can have very bad consequences. The normal phenylalanine level in the blood is about 3 to 5 mg per 100 ml, but in PKU this rises to

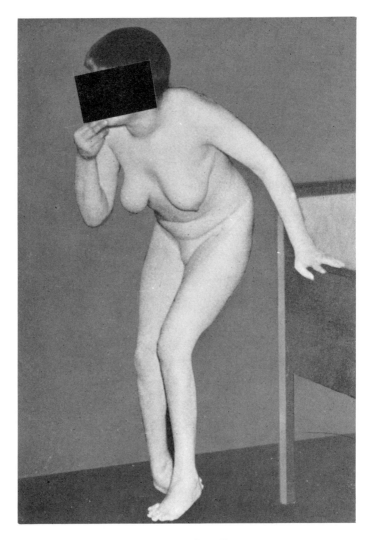

FIGURE 9–5

A thirty-year-old woman with PKU. The excess of phenylalanine and its alternate product phenylpy-ruvic acid has caused extreme mental deficiency and prevented the normal development of her legs. (Photograph courtesy of Carl Larson, University of Lund, Sweden)

80 even with the kidneys working at maximum capacity to remove it. Since the conversion is made in the liver, an embryo who is homozygous for the recessive gene will be normal as his mother's enzyme keeps the level normal. At birth, however, the level begins to go up. The sad part of this is that the excess phenylalanine interferes with the development of the brain, and the child grows up to be defective mentally. About one percent of those confined in mental institutions are there because of PKU. Also, the bones of the legs may not develop normally, so the child never learns to walk.

Much of the damage can be prevented if the baby with PKU is put on a diet very low in phenylalanine. This is done by preparing a milk of mixed amino acids leaving out this one. The powdered amino acids, **Lofenalac,** are mixed with water for use. Certain fruits and vegetables as well as carbohydrate foods can be given, since these are low in amino acids. Early recognition of the condition is essential. If it is detected only when the child shows that he is not developing normally, some irreversible damage to the brain has already taken place. Even the urine test does not become positive early enough. Now, however, the Guthrie test can show the condition as early as three days after birth. A drop of the baby's blood is placed on an absorbent paper. This is placed on a culture of bacteria which need phenylalanine. A heavy growth of bacteria around the paper shows that this amino acid is too high in the baby's blood. Most states now have laws which require that the Guthrie test be given to all newborn babies. Even though only about one in ten thousand will be positive, this one can be saved from a life of idiocy when the condition is recognized early.

The diet must be continued for about eight years; then the brain is well formed enough that it does not seem to be damaged by high phenylalanine in the blood. Should the person be a female, however, she must return to the diet if she ever becomes pregnant. The excess amino acid in her blood will be absorbed by the embryo and cause brain damage before birth if this precaution is not taken.

Carriers of the gene for PKU can be recognized by having them drink water with dissolved phenylalanine and checking the level of this amino acid in the blood over a period of about twelve hours. In those homozygous for the normal gene, the level quickly drops to normal, within two or three hours. In carriers, about twelve hours are required for it to return to normal.

Other Breaks In The Chain.   One enzyme break occurs in the chain of conversions leading to the production of the thyroid hormones, as has already been discussed. Another occurs in the chain leading to the formation of melanin. A break in this chain results in albinism, which was discussed in Chapter 4. As of now, nothing can be done to correct albinism. It is due to the lack of a product within the cells, and there is no way we can introduce the product into the cells, nor can we put the missing enzyme into the cells. Most of the tyrosine goes along another pathway which breaks it down into carbon dioxide, water, and simple nitrogen compounds. A break in this chain can lead to a build up of tyrosine in the blood and its excretion in the urine. **Tyrosinosis** is the result. This seems to do no great damage to the body. One of the later products in this chain is known as **alkapton,** and a break just below this point results in **alkaptonuria.** There is an excess of this compound in the urine. Alkapton turns a dark color when exposed to light, so a mother can diagnosis the condition easily by the dark discolorations on soiled diapers. No great harm comes from this excess in the blood although in later life the accumulation and darkening of alkapton in the cartilages will cause them to be dark. This shows primarily in the ears and the front of the nose where the cartilage is only covdered by skin. The accumulation may also be associated with arthritis.

Chapter 10

WHEN CHROMOSOMES MISBEHAVE

CHROMOSOME RESEARCH

The more we learn about human chromosomes, the more we realize the importance of a proper balance of these vital carriers of the genes. The loss or an addition of a chromosome, or even a small piece of one, can greatly influence development. Our knowledge of human chromosomes is comparatively new. Back in the twenties, T. H. Painter, at the University of Texas, made studies of the chromosomes as he saw them in thin sections of the tubules of the testes. He concluded that man's chromosome number was forty-eight, and this was accepted for years. The chromosomes in these cells were so small, however, that it was not possible to be sure of this number. Then, J. H. Tjio, studying in Sweden, found a way to see human chromosomes much more clearly. He took small bits of tissue from young aborted embryos and squashed them flat on a microscope slide. The cells in such tissue are growing rapidly, and many of them are in stages of divisions when the chromosomes show most clearly. All of the chromosomes of a cell thus become visible in one field and are larger and more easily counted than is possible in the thin sections of cells from the testes. Tjio found that there were only forty-six chromosomes in each cell. He reported his findings at the International Congress of Human Genetics in

Copenhagen in 1956. Geneticists from all over the world then began intensive study of human chromosomes.

One center of study was at the University of Colorado Medical School in Denver. There, Theodore Puck, a biophysicist by training, assembled a team of researchers, including Tjio, and worked out the exact nature of each chromosome. At a meeting in Denver in 1961, the Denver system was accepted as a standard. This system gives a number to each chromosome pair, and each group of similar chromosomes is given a letter designation. The length of the chromosome and the point of attachment of the centromere are considered. The chromosomes studied are usually in the prophase of mitosis, and each chromosome is already double. The centromere holds the two chromatids together. Table 10–1 shows the Denver system.

TABLE 10–1

Human Chromosome Designations
According to the Denver System

Group Designation	Chromosome Numbers	Characteristics of the Chromosomes
A	1–3	very long with median centromeres
B	4–5	long with submedian centromeres
C	6–12	medium length with submedian centromeres
D	13–15	medium length with near terminal centromeres
E	16–18	short with median centromere in 16 and submedian in 17 and 18
F	19–20	shorter than E group and with median centromeres
G	21–22	very short with submedian centromeres
Sex Chromosomes	23	X similar to C group, Y similar to G group, but the chromatids tend to remain closer together

How A Karyotype Is Made. The early work on embryo cells was very revealing, but geneticists needed a way to study the chromosomes of human beings after birth. One way was found; a small piece of skin was cut off, chopped up, and placed in a nutrient medium. Within two or three weeks there was some growth and cell division, and chromosomes could be seen in squash preparations. Then it was found that a little bone marrow taken from the breast bone (sternum) contained many dividing cells, and chromosomes could be seen in these without culturing the cells. This was not too popular, however, because it involved insertion of a stiff needle into the bone. The next step was made possible by the discovery that an extract from kidney beans would do two things to human blood. It would clump the red blood cells and stimulate some of the white blood cells into growth and division. The extract is known as **PHA (phytohemagglutin).** By this means, several drops of blood can be treated and placed in a nutrient solution, and within three days the dividing cells can be squashed on a microscope slide for study.

To make a karyotype, a photograph is made of the squashed cell and an enlarged print made from the negative. Then, with a sharp pair of scissors, each chromosome is cut out and matched up in pairs according to number and affixed to a sheet. This is rather costly at present because it requires considerable time of an experienced person. Tests are being made to turn the job over to computers. The chromosome smear is photographed, and a print is fed into a computer which then identifies each chromosome and indicates any which are abnormal. This would be a great time saver and may make possible many more studies of karyotypes to detect abnormalities.

ONE CHROMOSOME TOO MANY

Mongolism is the common name given to a type of abnormality which appears in about one in each seven hundred births in the United States. Children of this type have a short stature, a slanting fold of the skin of the upper eyelids, broad hands with stubby

FIGURE 10–1

A boy with Down's syndrome, mongolism, or trisomy-21. This boy is receiving special training which will enable him to achieve his maximum potential within the limits imposed on him because he has one chromosome too many.

fingers, a wide rounded face, a large tongue which may be furrowed, and are subnormal mentally. Many theories have been proposed to account for this type of abnormality. There was one theory in Europe during the last century that the Caucasian race developed as an offshoot of the Mongolian race. When these abnormal children appeared, the hypothesis was proposed that the embryos were arrested in their development and thus showed the Mongolian characteristics of their ancient ancestors. (The slanting fold of the eyelid was the similarity to the Mongolian race.) This was the origin of the term mongolism. Actually, the condition is found in all races including Mongolians. These people feel that the afflicted children look like Caucasians. A man in England, Langdon Down, made an extensive study of children with mongolism, and the name **Down's syndrome** is also used to designate the condition.

It has been found that the chances of woman having a child with this syndrome increases with age. About half of such births in the United States are to mothers over forty, yet only about 4 percent of the total births come from such women. This finding led to the theory that some degenerative changes in the uterus might interfere with normal development and Down's syndrome would result. When it became possible to see human chromosomes clearly, the true explanation was found. Children with the syndrome had forty-seven chromosomes; one too many. Chromosome-21 was present in triplicate. All the genes might be

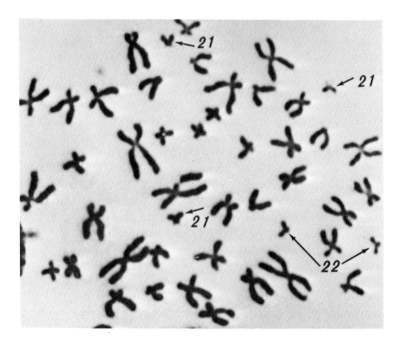

FIGURE 10–2

Chromosomes of a person with Down's syndrome. The arrows indicate the three chromosome-21's. Chromosome-22's are also shown for comparison.

normal, but three sets of genes on this very short chromosome together with only two sets of genes on the other chromosomes result in the syndrome. In Chapter 5 we learned how sexual deviations result from abnormal variations in the sex chromosomes, even though the coiled X in the Barr body makes possible an equalization of the active genes in both sexes.

Down's syndrome arises when there is non-disjunction of chromosome-21 during meiosis, so that one egg or one sperm received two 21's. When fertilized by a normal gamete from the opposite sex, the result will be three 21's. It is said to be **trisomy-21.** This is more likely to happen in egg formation than in sperm formation, and the chance of it increases with the age of the woman, but not the man. One theory is that the chromosomes pair long before the completion of meiosis, and the longer the paired chromosomes stay together, the greater the chance that they will not separate at meiosis. As a woman grows older, she has been exposed longer to natural radiation and perhaps other agents which are known to increase the chance for non-disjunction of chromosomes. While the average incidence of Down's syndrome in the United States is about one in six hundred, a woman in her twenties has only about one chance in fifteen hundred of having such a child at any one birth. If she waits until she is past forty, however, the chance jumps to about one in eighty, and beyond forty-five it is one in forty-four.

Children with Down's syndrome have always tended to die young. They are very susceptible to respiratory infections, and in the past few survived beyond twenty years of age. Today with modern antibiotics, this age is being extended. Also, in former years it was customary to merely take care of such persons, but now many places give them training within their capabilities, and they can hold jobs and earn their keep in society.

Is Mongolism Inherited?   When a married couple have had one child with mongolism, they face a crucial decision. Should they have any additional children? What is the chance of the same abnormality appearing in possible future children? Non-disjunction seems to be a random occurrence, and this knowledge has led many physicians to tell a patient that the chance for a second child of this type is no more than for any other woman in her age group. In most cases this has proved to be sound advice, but in some a second or even a third child with Down's syndrome has shown that there can be a transmissable factor in some families. Now we know why.

A chromosome study of many children with Down's syndrome shows that about 5 percent of them have the normal forty-six chromosomes. When these are examined more closely, however, it can be seen that one of the longer chromosomes has a small chromosome-21 attached to it. This is known as a **translocation.** Such children thus have three sets of genes on the chromosome-21. Examination of the parents of such children

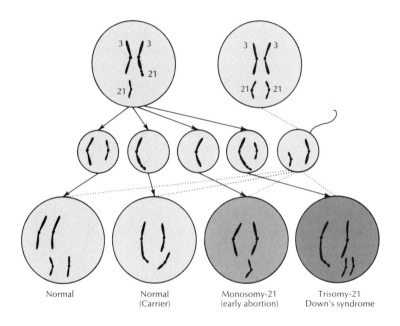

| Normal | Normal (Carrier) | Monosomy-21 (early abortion) | Trisomy-21 Down's syndrome |

FIGURE 10–3

How Down's syndrome may be transmitted from normal parents because of a translocation of chromosome-21 onto one of the longer chromosomes.

will reveal one with only forty-five chromosomes. There will be one free 21 and one 21 attached to a longer chromosome. They have a normal set of two genes of each kind. Some of their reproductive cells will carry the free 21 together with the attached

21. A child formed from such a reproductive cell will be a forty-six chromosome mongoloid. Other children will receive only the attached 21 and will be normal, but can have trisomy-21 children. Hence, it is only common sense for any parents who have had one such child to have karyotypes made to see if a translocation is involved before they consider having more children. If so, only half of future conceptions will be normal and half of these will be carriers. One-fourth will be mongoloid, and one-fourth will be early abortions. This one-fourth will have only one 21, and this has a lethal effect.

When a younger woman has a child with mongolism, there is a greater chance that it will be due to translocation than in older women. This is because the chance of translocation remains constant regardless of age, but the chance of non-disjunction increases with maternal age. Hence, while the actual number remains the same, the proportion due to translocation decreases with maternal age.

What about the zygotes which receive only one 21? Both in translocation and in non-disjunction, each time you get a trisomy-21 you also get a monosomy-21. Although such zygotes may have all the genes needed to form a body, the fact that there is only one set of genes on this short chromosome so greatly upsets the gene balance that the embryos live only a short time, and there is early abortion.

NON-DISJUNCTION OF OTHER CHROMOSOMES

We have learned that non-disjunction takes place with regard to the sex chromosomes and also with the chromosome-21's. What about the other chromosomes; do they also have non-disjunction? The answer is yes, but the abnormalities resulting from the trisomic condition of the other chromosomes is so extreme that they rarely survive their intrauterine existence. Some have been found, however.

Trisomy-18 (Edward's syndrome) is found in about one live birth out of every forty-five hundred. These babies have low-set ears, a rounded bottom of the foot, heart and kidney defects, and survive only about ten weeks on the average.

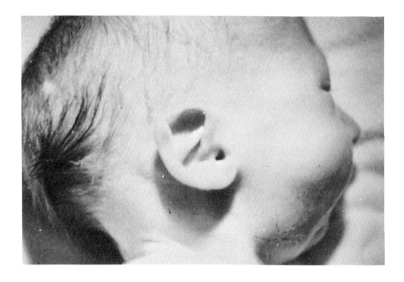

FIGURE 10-4

Head of a baby with trisomy-18 showing the peculiar ear that characterizes this condition.

Trisomy-13 (Patau's syndrome) has an occurrence of only one in every 14,500 live births. These children have broad noses, small craniums, small eyes and are usually blind, and have heart abnormalities. They live only a few weeks.

Trisomy-22 is very rare, but has been found. Apparently this little chromosome has some very influential genes and causes a great upset of the balance when it is present in triplicate. It is an exception because, as a general rule, the longer the chromosome which is trisomic, the greater will be the effect. Studies on the chromosomes of spontaneously aborted embryos show that about 40 percent of them have chromosome abnormalities. Many of these are trisomics. The trisomy of the longer chromosomes are found in the embryos which are aborted early, and then those of median length in those which abort a little later, and finally the short ones for the late abortions. **Monosomics,** those having only one chromosome of any pair, seem to be so abnormal that the embryonic deaths are very early, and they are seldom seen.

CHROMOSOME BREAKS

Sometimes chromosomes may break at a point along their length. The portion broken off may be lost because a piece of a chromosome without a centromere to guide it in cell division gets left out of the newly forming nuclei and disintegrates in the cytoplasm. This results in a **deletion,** the loss of a part of a chromosome. At other times the part which breaks off may become attached to another chromosome, and we say a **translocation** has occurred.

The Philadelphia Chromosome, A Somatic Deletion. Chronic **granulocytic leukemia** is characterized by an excess of those leukocytes (white blood cells) with cytoplasmic granules. One study of a karyotype made from cells taken from the bone marrow of a child who was suffering from this type of leukemia

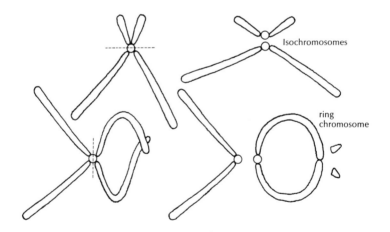

FIGURE 10–5

How isochromosomes and ring chromosomes may be formed. A horizontal cleavage of the centromere results in two chromosomes with equal halves. A double deletion and attachment of the broken ends gives a chromosome with a ring shape.

showed one chromosome-21 which was shorter than the other. About 29 percent of the chromosome was missing. Since this study was made in Philadelphia, this became known as the Philadelphia chromosome, and has since been found to be associated with this type of leukemia in studies made all over the world. The strange thing about it is that karyotypes made from tissue culture cells from other parts of the body do not show this short chromosome. Evidently, there is a weak place at a certain point on this chromosome which causes it to break easily with a resulting deletion. Some environmental factor, such as virus disease, radiation, or something else may trigger such breaks. It is possible that only the cells in the bone marrow, where the leukocytes are produced, are susceptible to these environmental inducers. Another possibility is that it may occur in different kinds of cells, but can survive and reproduce only in the bone marrow cells.

The Cry Of The Cat. Another unusual abnormality results from a deletion of a part of the short arm of chromosome-5, and this one is found in cells from all parts of the body. The vocal cords fail to close at the bottom, causing an afflicted baby to cry with a sound that is similar to a cat mewing. The French first named it, *cri du chat,* cry of the cat, syndrome. Other features include a beak-like profile and mental retardation.

Unusual Chromosomes Resulting From Breaks. Sometimes deletions may occur at one or both ends of a chromosome, and the remainder joins to form a ring chromosome, Figure 10–5. Isochromosomes are formed when there is a vertical division of the centromere during cell division, as seen in Figure 10–5. Cancer cells usually show abnormal chromosomes, and nearly always chromosome-16 is involved. This chromosome must have genes which correlate the growth and differentiation of cells. Practically all of the chromosome aberrations which cause an upset of the gene balance result in abnormalities. It is estimated that about 2 percent of all conceptions include some type of chromosome aberration. Fortunately, the great majority of these have a lethal effect and bring about early abortion. The ones we have been describing are the rarer ones which are relatively mild in their effects and which cause some of the embryos to be normal enough to survive to a live birth. Even among these the majority are aborted before time for birth.

INDUCED CHROMOSOME ABERRATION

The same agents, as a whole, which cause gene mutation also cause chromosome aberrations (see Chapter 5). High-energy radiation has been shown to cause both non-disjunction and chromosome breaks. LSD was first tested on cells growing in tissue cultures and was found to greatly increase the number of chromosome breakages. Then blood cultures were made from persons who had used LSD in considerable quantities over a long period of time. The results from a number of different studies were at first conflicting. Some reported positive results while others reported negative results. Possibly differences in the selection of subjects were involved, as well as differences in the staff making the studies. In some laboratories where great care

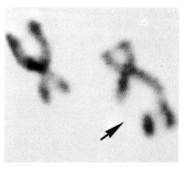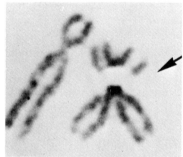

FIGURE 10–6

Chromosome breakage induced by LSD. These chromosomes were found in blood tissue culture cells from a man who had used LSD rather heavily for several years. About 20 percent of the cells showed such breakage as compared with about one percent in control cells from a person who had not used the drug. (Photograph courtesy of J. Egozcue)

was taken to select persons who really had been taking LSD extensively and where the laboratory staff was experienced, as many as 25 percent of the cultured cells were found to contain one or more chromosome breaks. Control studies showed less than 2 percent breaks.

The virus diseases also can produce breaks in chromosomes. These are probably involved in the many abnormalities which arise in an embryo when the mother has such a disease during early pregnancy. Among those diseases which have been shown to have this effect are: rubella (german measles), measles, smallpox, chickenpox, mumps, Herpes II (shingles), and infectious hepatitis. Influenza is also highly suspected. Extracts of these viruses have been shown to cause breaks in tissue culture cells.

Chapter 11

HEREDITY AND ENVIRONMENT

Is hay fever a result of heredity or environment? Many people suffer from the sneezing, copious nasal secretions, and irritated eyes that typically appear at certain seasons of the year. If a victim of such symptoms goes to a physician, he is likely to be given some skin tests and may be informed that he is allergic to a certain weed pollen. Whenever this pollen is abundant in the air, there is an antigen-antibody reaction in the nasal cavity and the mucous membranes of the eyes. The hay fever symptoms result.

Hence, it might seem reasonable to conclude that this is purely an environmentally induced reaction. If this is true, however, why do most people escape hay fever symptoms? In the city of Indianapolis, for instance, on about August 15 each year, perhaps as many as 10 percent of the city's population begin the annual inconvenience of ragweed hay fever. The other 90 percent inhale the same kind of pollen, but they do not have hay fever. Heredity is at least a partial answer. A study of many families showed that there is an inherited tendency to easy sensitization against foreign proteins. A man may have ragweed hay fever and two of his four children may develop it also. They did not inherit ragweed hay fever, however. Had the family moved to Hawaii, none of the children would have been sensitive to this pollen. The two children might have become allergic to certain foods or other items in the environment because they inherited the ability to become easily sensitized, but it is only after contact with the offending protein that an allergy can develop.

This is a good illustration of the relationship between heredity and environment. A person inherits a potential, but environment is necessary for the attainment of the potential. A child who inherits the genes for PKU, does not inherit an idiot mentality. He merely inherits an inability to breakdown a certain amino acid (see Chapter 9). It is the build up of the amino acid and its alternate by-products which damages the brain and results in the idiocy. By removing the amino acid from the diet we can prevent the brain damage. Likewise, a cretin does not inherit a retarded physical and mental development. He inherits an inability to produce an enzyme needed to synthesize thyroid hormones. By giving him the hormones, we can make him normal. Thus, we should think of a person's total development as a result of environmental forces acting upon a genetic potential.

The exposure of a photographic film and its subsequent development can be correlated with heredity and environment. Suppose you load your camera with color film and make some exposures. You have no image on the film; the exposed and unexposed portions will look just the same. You have created a potential picture, however. The light striking the film has activated silver compounds so that they are now responsive to chemical development agents. This is the equivalent of heredity. The development of the film is equal to the environment. By applying the proper chemical solutions at the proper temperature for the proper time, the potential image becomes an actual image, and a color transparency results. You may have made a perfect exposure, yet if the film has poor development, the result will be poor. Likewise, if you do not make the proper exposure, then normal developing will not result in a good picture. If the processor knows that you have made a mistake in exposure, say you have underexposed, he can make certain compensations in development and perhaps get a satisfactory picture. This is similar to those cases where we can correct known genetic deficiencies by compensating environmental controls.

In this chapter we will try to get a clearer picture of the role of heredity and environment and how each is related to the other.

ENVIRONMENTAL MIMICS OF GENETIC TRAITS

The color of the skin is one of the distinctive characteristics of the human races. Natural melanin deposits vary from the fair-skinned Nordics to the black Africans, and no one would question the genetic basis for these skin-color differences. Yet we also know that exposure to sunlight affects the amount of melanin in the skin. Consider a Hawaiian boy. He inherits a medium amount of melanin deposit and will have a nice bronze skin color regardless of whether he ever gets in the sun. A fair-skinned Norwegian who moved to Hawaii may enjoy the swimming and surfing so much that he spends much time on the beach and soon develops such a deep tan that his skin will actually be as dark as the Hawaiians. Thus, environment has duplicated the skin trait which is inherited by Hawaiians. The Norwegian has inherited an ability to respond to sunlight by heavier deposits of melanin. Such environmental duplication of genetic traits are called **phenocopies.**

For many years physicians throughout the world have known of relatively rare cases of babies born with arms and/or legs greatly reduced in size and defective in shape. The condition was called **phocomelia,** and pedigree studies showed that it was due to a recessive gene. Then, in Europe in 1961, the incidence of babies with this abnormality showed an alarming increase. As brought out in Chapter 3, this was because of the use of thallidomide by many expectant mothers. This drug was doing the same thing to the developing embryos as genes had been doing, preventing normal growth of one or more of the limbs. Thus, the drug was resulting in a phenocopy of what had been considered a purely genetic trait. Hence, geneticists and physicians must use great care in classifying any trait as being due primarily to heredity or environment. A particular trait may be due primarily to heredity in some cases and to environment in other cases.

Another good example to illustrate this point is **hydroceph-alus,** sometimes known as water on the brain. A child may be born with a greatly swollen cranium, due to internal pressure on the soft bones of the developing embryo. Within the normal brain there are cavities, **ventricles,** filled with a cerebrospinal fluid under a specific pressure. This pressure is needed to hold the soft brain out against the skull. The pressure is regulated within a narrow tolerance. If it goes up a little, then some of the fluid will be absorbed through the membrane around the ventricles. Also, there is an escape valve through a duct down into the spinal column. One autosomal recessive gene is known which causes a thickening of the membrane around the ventricles, and this causes such a build up of pressure that the brain expands outward causing the bulging forehead characteristic of hydrocephalus. A sex-linked recessive gene causes an obstruction of the duct, and hydrocephalus also develops. Hence, if one asks if hydrocephalus is inherited, one might arbitrarily answer, yes, as a result of either one of these genes. You could cite family pedigrees to support your answer. Such an answer would not consider the possibilities of phenocopy. Some cases are known in which a pregnant woman had a virus infection when the membranes of the ventricles were forming causing them to become thickened and hydrocephalus resulted. Other cases have shown that heavy radiation at a certain time can cause imperfect development of the duct from the ventricles, and again hydrocephalus results.

We can now prevent this condition. A tube can be inserted into the ventricles at birth and, with a valve control, can carry the excess fluid down through the neck and chest to drain into the abdomen. Thus, we can produce a phenocopy of a normal child even though the genes may code hydrocephalus.

In experimental animals we can produce phenocopies of many genetic defects. Monkeys have been produced with all the symptoms of PKU by giving them much greater than normal amounts of phenylalanine in their diets. They inherited genes which could produce enough enzymes to process normal quantities of this amino acid, but it became too high in the blood when the intake was so great. Rats have been induced to show all the symptoms of galactosemia by feeding them high galactose diets. Such studies give us a better understanding of the relative roles of genes and environment.

VARIABLE EXPRESSIVITY AND REDUCED PENETRANCE

Genes can vary in the degree of their expression as a result of their interaction with the environment. Some persons inherit the genes for **gout,** but if they are careful of their diet and take the proper medication, they may only occasionally show rather mild symptoms. Another person with the same genes, however,

FIGURE 11-1

Variable expressivity of a dominant gene for flipper arms. The father, on the right, does not have such an extreme shortening of the arms as his son, on the left. Modifying genes or some factor in the environment of the developing embryos can alter the degree of expression of some genes such as this one. (Photograph courtesy of Karl Stiles)

may find the temptation to enjoy a thick beefsteak too great and may suffer excruciating pain from the crystals which form in his joints. This is variable expressivity of genes as a result of variable environmental factors.

Patients with PKU in mental hospitals may show variations all the way from a low level idiot up to the moron level. This might be explained by variations in protein intake in early life plus possible other genes which modified the effect of the phenylalanine on the developing brain.

Sometimes it is difficult to find an external environmental factor which might be involved, and the variations seem to be due entirely to other genes within the cells. **Blue sclera,** for instance, is caused by a dominant gene. In a few of those who inherit this gene the sclera (white) of the eyes is such a dark blue as to appear almost black. Others with the same gene will show varying shades of blue all the way to such a light blue that it cannot be distinguished from the normal sclera. About 10 percent of those afflicted have this very light blue deposit, so we say that this gene has **incomplete penetrance.** Those 10 percent carry the dominant gene but show no detectable signs of it. These persons, however, may have some children with the very dark blue sclera. Also, the gene results in brittle bones in about 75 percent of those who carry it. In some of these the bones are only a little more brittle than normal, but at the other extreme the bones may be so brittle as to be broken with the slightest strain. A break can result from catching the leg on a sheet while turning over in bed at night. Modifying genes seem to explain these variations in penetrance and expressivity.

Thus, contrary to the general rule, there are cases where a person can carry a dominant gene without expressing it and yet pass the trait on to half of his children.

WHAT TWINS TELL US

The study of heredity and environment is greatly aided by the fact that two different kinds of twins are available for study. Since identical twins have identical genes, any variations they show must be due to environment. These can be compared to

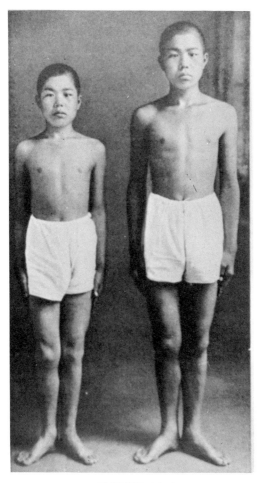

FIGURE 11–2

Environmental alteration of genetic potential. These identical twins have identical genes for body stature, yet one is much taller. The one on the left had a severe infection at five years of age. This hindered the development of the part of his pituitary gland which produces the growth hormone. As a result, he has not been able to attain as much of his genetic growth potential as was possible for his twin brother. (Photograph courtesy of Taku Komai and G. Fukoaka)

fraternal twins who will show differences due to both heredity and environment. For such comparisons we consider only the same-sexed fraternal twins, because sexual differences could confuse the conclusions. Also, we sometimes consider same-sexed brothers or sisters born at different times. These, known as **sibs,** will have the same genetic differences as fraternal twins, but because of being reared at a different time, there could be greater environmental differences. The most valuable results have come from identical twins reared apart. There are not too many of these, but enough have been located to show some interesting results. With identical genes, separated at birth and raised by different foster parents, often in different parts of the country, the influence of a different environment has a maximum opportunity of showing its effect.

Concordance is used in comparing the similarities and differences in the different kinds of twins. This term simply expresses the percentage of similarity with respect to the presence of a trait. A high concordance in identical twins with a much lower concordance in fraternal twins would indicate a very strong influence of heredity. A concordance which is very much the same between the two types of twins would indicate a predominantly environmental influence.

As an example, a study of albinism in identical twins shows a concordance of 100 percent. This means that whenever one is an albino, his twin is also albino in all cases studied. A study of fraternal twins, however, shows a concordance of only 25 percent. When one is an albino, the other is albino in only one-fourth of the cases. This would be expected since the chance of any one fertilization resulting in an albino is one-fourth. These results indicate that albinism is a result of heredity and that environmental factors do not modify it.

In a similar study of **harelip,** incomplete fusion of the upper lip, the concordance for identical twins was 33 percent and only 5 percent for fraternals. Such a low percentage for identicals indicates a strong environmental influence which can affect one twin without affecting the other in the same uterus. The difference between the two types of twins, however, shows that there are some genetic factors involved. (Studies on mice indicate that this trait may be genetic in some instances and environmental in others.)

In cases of deafness resulting from rubella infection of the mother a concordance of 86 and 88 percent was found for the two types of twins. Such a close agreement would indicate that this is a trait which is altogether due to environment. The fact that there is a discordance of 14 and 12 percent might be explained by the fact that there can be small differences in the time of ear development in a pair of twins.

For traits which differ quantitatively we generally study the twin differences in actual measure. Table 11–1 shows a few such traits. These figures indicate that body stature is very strongly influenced by heredity. Body weight is also influenced, but less than height.

TABLE 11–1

Differences in Twins and Sibs with Regard to Some Quantitative Traits

Traits	Identical	Fraternal	Sibs	Identicals Reared Apart
Stature (diff. in cm.)	1.7	4.4	4.5	1.8
Weight (diff. in lbs.)	4.1	10.0	10.4	9.9
Age of first menstruation of girls (diff. in months)	2.8	12.0	12.9	0
I. Q. (Binet) (diff. in points)	5.9	9.9	9.8	8.2

HEREDITY AND DISEASE

Disease is such a broad term that it can refer to almost any deviation from the normal state. For convenience we can divide it into two classifications according to cause. **Infectious diseases** are those caused by an invasion of the body by microorganisms, such as bacteria, viruses, or protozoans. **Non-infectious diseases** are those caused by organic failures or deficiencies of food elements, vitamins, or hormones. Table 11–2 shows concordance for a number of human traits including a number of diseases.

TABLE 11–2

Concordance (Percent Similarity) for the Two Types of Twins
With Respect to Certain Human Characteristics

Trait	Monozygotic Twins	Dizygotic Twins
Measles	95%	87%
Tuberculosis	65	25
Diabetes mellitus	84	37
Rickets	88	22
Cancer	61	44
Site of cancer (where both have cancer)	95	58
Clubfoot	32	3
Harelip	33	5
Down's syndrome (Mongolism)	89	6
Epilepsy	72	15
Schizophrenia	86	15
Manic-depressive psychosis	77	19
Mental retardation	97	37

At first thought one might assume that infectious diseases would be purely environmental since invading germs are from the external environment. The concordances, however, show otherwise. Since most Americans are susceptible to **measles,** the concordance is very high for both types of twins, so heredity would play little part. **Tuberculosis,** however, is a different story. The differences in concordances indicate that people can inherit a difference in susceptibility to this disease. This concept is borne out by the observed fact that there can be great differences in susceptibility among different races. The American Indians had never contacted the germs of tuberculosis, so there had been no selection for resistance to this disease in their ancestry. When white man came and brought the bacteria with him, there were many deaths from the disease. The entire population of some islands in the South Pacific were wiped out by these bacteria brought by English sailors. This shows that not only the chance of infection, but the severity of infection is strongly influenced by heredity. A tribe of Indians, the Timucans, living on the west coast of Florida was wiped out by the chickenpox virus introduced by English sailors.

Among the non-infectious diseases caused by food deficiencies, again environment might seem to be the primary cause, but the concordances show otherwise. **Rickets** is due to a deficiency of vitamin D in the diet. Yet heredity influences the quantity of this vitamin which is needed to prevent rickets. Selection in rats has established some strains which need several times as much vitamin D as other strains need to prevent the bone abnormalities.

Environmental factors can certainly induce **cancer**; high-energy radiation is one such agent. Yet twin studies show some difference in a person's chance of developing cancer according to his heredity.

Sugar diabetes has long been a medical puzzle. A young woman, for instance, may have been normal all her life, but in her early twenties, she becomes lethargic, especially after meals, and may lapse into a coma. A physician diagnoses the trouble as diabetes when he finds sugar in her urine. He says that she has insufficient insulin, a hormone which regulates the absorption of glucose by the cells. When the young woman begins taking insulin and carefully regulates her carbohydrate intake, she becomes her old self again. Attempts to find a simple single gene method of inheritance ran into difficulties, yet heredity must be involved as shown by the twin studies. In some cases it seems as if a person inherits overactive pancreatic cells which secrete too much insulin during childhood, creating a great desire for sweets. The body may respond to this by producing an insulin antagonist which in time may overdo its job and reduce the active insulin below the required level. Then the symptoms of diabetes appear. In other cases other genes may be involved. Some may inherit kidneys which fail to respond to the excess of glucose in the blood and do not remove it until the level is too high. Still other genes can interfere with absorption of glucose at the cellular level. Environment is certainly a factor, because some who have the inherited tendency may escape if they have a diet which is moderate in carbohydrates. There was one case of identical twins. One became a tavern owner and became a heavy drinker of beer with its high carbohydrate content and developed diabetes. His twin, with the same inherited potential, had a much lower carbohydrate intake and escaped.

HEREDITY AND MENTAL DEFECTS

Man is superior to lower forms of animal life only with respect to his high degree of intellectual development. Anything which lowers his intellectual potential, therefore, makes him that much closer to life on the level of lower animals. There are so many things which can bring about a decrease in this potential. Purely environmental factors may be involved. A baby's brain injured in a difficult birth may never reach the intellectual development that we call normal. An adult man who receives a very hard blow on the head may lose much of the intelligence he has already developed. Disease germs, such as those of syphilis may invade the brain and bring about irreversible damage at any age of life. Even a lack of normal training of the intellect can prevent normal attainment of mental abilities. Perhaps as many as one-third of the total cases of mental defects can be traced to purely environmental causes. For the other two-thirds, heredity is involved in varying degrees.

The brain is such a delicate organ that many different kinds of changes in body metabolism and growth can result in a reduction of its potential. Enzyme deficiencies which result in too much of an amino acid or too little of a product can prevent normal brain growth and function. Chromosome aberrations which upset the balance of genes with one another produce many different physical syndromes, but mental deficiency is very often an accompaniment of these abnormalities.

The one mental defect which is most common is known as **schizophrenia** or split personality. It takes different forms and is found in different degrees of severity. In some persons it is expressed as **paranoia,** a persecution complex. In others it may be a **catatonic** withdrawal, where a person is oblivious to his surroundings. In still others there may be **delusions of grandeur**

or just plain silliness. About 1 percent of our people are potential schizophrenics and about half those in mental institutions
suffer from this mental defect. It may come on at different times
of life, but about 75 percent of the cases develop before age 25.

Many psychiatrists in the past felt that it was purely environmentally induced. They observed many cases where normal persons underwent a great emotional strain, and this mental
aberration developed as a result. Still, the twin studies which
showed concordances of about 86 and 15 percent for the two
types of twins indicated that there must be some predisposing
hereditary background. A better understanding of the part
played by heredity and environment has resulted from chemical
studies of the brain. A certain chemical, **serotonin,** is needed for
normal brain function. Under conditions of extreme mental
stress, all persons react with a change in the hormonal and chemical balance of the blood. Some inherit genes which release a
chemical under stress which stimulates a release of an excess of
serotonin. This is what brings about the halucinations and other
symptoms of schizophrenia. Persons without the genes for the
production of this chemical stimulant can withstand great
amounts of stress without mental breakdown. Certain halucino-
genic drugs, such as LSD and mescaline, seem to work on the
brain in a similar way. They can induce temporary schizophrenia.
Our problem is to find a drug which will bring the level down to
nomal. If this is done, we may be able to conquer this great human affliction.

Mental Retardation.   Intelligence is a trait which shows a continuous variation from very low to very high with most persons
falling within twenty points of the average as measured by I.Q.
scores. Such a distribution is typical of polygenic inheritance.
Many genes are involved in the development of the brain and its
complex process of functioning. Variations in these genes can
result in variations in potential intellect. Environment plays a
great part in the degree of expression of many of these genes.
One study of thirty-five year old identical twins showed an I.Q.
difference of twenty-four points. They had been separated when
still babies and raised in very different educational environments. One had dropped out of school after finishing the second

grade while the other continued on through college. This is an extreme case which illustrates the importance of training if the genetic potential is to come close to being realized.

Somewhere along the continuous scale of mental ability we draw a line and say that all below this level are mentally retarded. Children in this group find it very difficult to compete with those of a higher potential in school. Today many special education programs are available so that such children can be trained at their level and many can become productive members of society. Even mongoloids which typically lie on the idiot or imbecile level of the I.Q. range can be trained to a level of accomplishment where they can perform useful services in society. Instead of spending their lives in inactivity, being supported by the state, they may work and earn their keep. Such programs have been observed in such widely separated areas as the Soviet Union, Australia, and the United States.

Chapter 12

GENES IN POPULATIONS

FORCES OF SELECTION

One does not require any particularly keen powers of observation to know that gene distribution varies among the different populations of the earth. Facial features, body build, skin pigmentation, and other inherited characteristics are all so distinctive that it is easy to recognize the major races. Then, there are variations within the races.

All people of the Mongolian race have distinctive features, but there are also distinctions between different national groups within this race. The Japanese, Koreans, Chinese, Vietnamese, and Cambodians all have their own particular features. How have these variations in gene distribution come about? Our best evidence indicates that man had his origin in Central Africa and then spread all over the earth. What forces caused the establishment of different kinds of genes in the different environments to which he migrated? At least three forces were involved.

Natural Selection. The forces of nature are constantly acting upon all living things, eliminating those less fit and establishing those best adapted to the environment in which they live. Most of these features of fitness are inherited, so various gene combinations become established, and these will vary according to the environment. To illustrate how this might have affected a human trait, let us consider the intensity of skin pigmentation. As a general rule, we find that the degree of melanin deposit is related to the intensity of exposure to the ultraviolet rays of the sun. Generally, those living near the equator have darker skins than those in the temperate zones. The heaviest melanin deposits are found in those who lived in the more open and exposed regions, such as those found in much of Central Africa and the interior regions of Australia. The natives of equatorial South America have less heavy deposits. There is a heavier rainfall in this area and more trees which provide shade, so less of the sun's rays penetrated to the people. As we move north from Central Africa and up through Europe we tend to find a gradually declining degree of pigmentation up to the fair-skinned Nordic populations. As we go still further north we again find an increase in pigmentation among the Laplanders. In the northern part of the Scandinavian peninsula there is a snow cover on the ground much of the year and a great reflection of the sun's rays. Those who have been sunburned on the ski slopes can testify to the importance of this reflection.

At first one might think that the burning of the skin was the primary factor in establishing these skin pigment differences. Perhaps, in a tropical climate, those with fair skins became severely sunburned and infections of the blistered skin might have eliminated them. Another factor, however, might be of even greater importance. Vitamin D is essential to man, but within critical limits. Rickets and bone deterioration can result when the quantity is insufficient. However, too much of this vitamin is harmful and can also result in bone defects. In fact, some persons have even died from too much of this vitamin; a hungry group of hunters in Alaska gorged themselves on polar bear liver, which is very rich in vitamin D, and several died as a result. This is a vitamin which is manufactured in the sunlight. We can get it by eating foods which have been in the sunlight, or we can manufacture it in our own skin when in the sunlight. It can be stored in the liver for use in periods when the sunlight is less intense.

The melanin in the skin acts as a filter which regulates the amounts of ultraviolet rays which reach the deeper skin cells. These cells can manufacture vitamin D, and can also be burned by too much exposure. Hence, selection favors just the right amount of pigmentation which keeps the vitamin D level at the optimum level. These are the people who are in the best health and who would reproduce themselves most abundantly. Selection has also made possible a temporary increase in pigmentation during seasons when the sun's rays are more intense, but this is followed by a return to fair skin which has a maximum chance of capturing the weak rays of a winter sun. After many thousands of generations of natural selection, each racial group arrived at an assortment of genes which gave the most favorable degree of pigmentation for the region in which it lived. Today with vitamin D additives in many foods and more controlled protection from the sun's rays, the selection pressure is greatly reduced, but people still reflect the genetic make-up of their distant ancestry.

Sexual Selection. The standards of sexual attractiveness may vary from one population to another, and this can be a factor in establishing gene distribution. Features of the face and body deemed attractive might cause some to mate earlier and more often and thus establish their genotype more abundantly in the group. Among the males, physical combat over possession of desirable females could lead to increased strength and prowess in the primitive societies. Even in the more advanced societies of today where most people find mates and reproduce, there are still many who are so far below the standards of sexual desirability that they never reproduce.

Genetic Drift. Chance sometimes plays a part in the establishment of the proportion of genes in different populations. It is especially important when a small group from one area migrates to another area and establishes a new population. The Maoris, natives of New Zealand, originated from a small group of Tahitians who set out to the South in their long canoes. The group who embarked on this arduous journey of about two thousand miles across open water was not very large, so they probably did not represent an exact proportion of the genes in the Tahitian population. As an example, they seemed to have had a much smaller percentage of the genes for type A blood. Then, when they reached New Zealand, it is quite possible that those who did have the gene for the A antigen did not have as many children as those who did not have this gene. As a result, we find that the Maories of today have a much smaller percentage of type A blood than is found in the Tahitians. They also vary in other inherited characteristics because of this influence of genetic drift. We, of course, must make allowance for possible mixing with other people during the long period since the migration and possible differential selection, but drift almost certainly is an important factor in explaining the differences in the two populations today.

Catastrophe may also play a part in genetic drift of established populations. N. E. Morton, of the University of Hawaii, studied the people on the island of Pingelap, one of the Micronesian islands in the South Pacific. He found that about 6 percent of them had a severe eye defect, **achromatophobia,** which results from the homozygous state of a recessive gene. Those who express this trait have abnormal cones in the retina. The cones are used primarily in bright light and are the bodies which are color sensitive. Hence, these people cannot stand exposure to bright light and are totally colorblind. Selection would seem to reduce the incidence of this gene since these persons could not get out and help collect food. It was found that around the turn of the century a great hurricane swept the island and killed most of the population. Only about twenty survived and these evidently, by chance, had a much higher than normal percentage of the gene for the eye defect.

Even when a small group migrates into an area populated by others, drift can play a part, provided the migrants do not intermarry with their neighbors. Bentley Glass studied an interesting population of Dunkers, a religious sect living in Franklin County, Pennsylvania. These are the descendants of a religious sect, the Baptist Brethren, who originated in the Rhineland region of Germany near Krefeld. In 1719, a group of twenty-eight of these migrated to America. They had strict rules against marriage with others outside the sect, so have remained isolated genetically. Some have tired of the strict religious restrictions and have migrated out, but there has been practically no migration from the surrounding area into the Dunker population. Here was an opportunity to study the possible effects of genetic drift though perhaps the original migrants were not an accurate representation of the group from which they came. Blood types were analyzed with the following results:

TABLE 12–1

Population	Percentage of ABO Blood Groups			
	O	A	B	AB
Dunkers	35.5	59.3	3.1	2.2
Rhineland Germans	40.7	44.6	10.0	4.7
Penn., U. S.	45.2	39.5	11.2	4.2

These results do show that genetic drift must have accounted for the higher percentage of A in the Dunkers than is found in the German people from which they descended. This could not be accounted for by an inflow of genes from surrounding Pennsylvanians because they have even less A antigens than are found in the Germans.

Selection Favoring Carriers Of Harmful Genes.   About 8 percent of the Americans with an ancestry from central Africa carry the gene for sickle-cell anemia. They have the sickle-cell trait as heterozygotes, but this trait does not greatly inconvenience them. It does lead to about one child in 170 with the dread sickle-cell anemia. Since those with the anemia almost never live to have children, it might seem that selection would long ago have reduced the incidence of this gene in the population. The gene is so rare among Americans of northern European ancestry that cases of sickle-cell anemia are almost unknown. How can we account for this difference? Any gene so prevalent as 8 percent must have had some advantage in the past.

A study of the incidence of the gene in world populations shows that it is most prevalent in central Africa, a region of India, and parts of Greece. These all happen to be areas which have had a high incidence of a very deadly type of malaria. Those persons who are heterozygous for the gene, however, seem to have a high degree of resistance to this disease. Hence, such persons have an advantage. Those who are homozygous for the gene for normal hemoglobin are likely to die of malaria, or at least be greatly incapacitated by it. Those homozygous for the gene for sickle-cell anemia will die early in life from anemia. This leaves the favored heterozygous persons as the primary procreators of the future generations. In one region of coastal West Africa the incidence of the gene in the population is as high as 31 percent.

Why should the carriers of this gene be more resistant to malaria? It seems that almost any gene which results in abnormal hemoglobin or other defects of the red blood cells, confers such resistance. **Thalassemia, Cooley's anemia,** is another example. In persons homozygous for this gene, the anemia is so severe that it is almost 100 percent fatal. The heterozygous person is highly resistant to malaria, so this gene has been established in regions of the world where malaria has been high. Italy is one such region. By chance this gene was selected for malaria resistance in these regions and the gene for sickle-cell anemia in other regions. The sex-linked gene, **G6PD deficiency** which gives a person severe anemia when he eats lima beans, also has value in malaria resistance for the heterozygous females, and its incidence is high in certain regions where malaria has been prevalent.

With these cases in mind geneticists began trying to explain the high incidence of other harmful genes in certain populations. The gene for **Tay-Sach's disease,** infantile amaurotic idiocy, is found in a high frequency in Jewish populations, but is almost unknown among other populations. (See Chapter 9 for details of this condition.) In New York City about one person in thirty of the Jewish population is a carrier of this gene. The normal allele of this gene produces an enzyme needed for processing the products of fats in the diet. When homozygous for the gene, the processing is faulty, and fatty accumulations appear around the nerve cells, gradually destroying them and bringing death within a few years. How could a carrier of this gene possibly have any survival advantage? A study of the carriers in New York City showed that most of them could trace their ancestry back to northeastern Poland and southern Lithuania. This was an area known in the past for the crowded conditions of the Jewish ghettos. The food available was far from adequate, especially in fats. Perhaps with two genes for the enzyme there was an over-action on the small amount of fat in the diet, and this was harmful. The heterozygous person, on the other hand, would produce less enzyme and could actually process his small amount of fat better. This is a condition known as **overdominance** which occurs when two dominant genes go too far in their actions, while one alone does a better job. Under these conditions, the heterozygous individuals would be better able to withstand life in the ghettos, and the gene for Tay-Sach's disease would be established at a high level in the population.

Another instance of great variation in the frequency of a very harmful gene was revealed by a study made by N. E. Morton in Hawaii. He found that cystic fibrosis had an incidence of twenty-six per one hundred thousand in the Caucasian population of the islands, while the non-Caucasians had an incidence of only 1.11 per one hundred thousand. Why should Caucasians have about twenty-five times more cases of the disease? We do not even have a speculative answer, but it is possible that Caucasians may have had an environment at one time that favored the heterozygous person.

EQUILIBRIUM OF GENE MUTANTS

All genes in a population tend to become established in a constant equilibrium as long as all environmental factors remain constant. Selection favors genes which have a beneficial phenotype, and these genes will be established at a very high level. Harmful genes will tend to be eliminated by genetic death. Genetic death simply means that a person expressing the gene does not reproduce himself. He may die in the embryonic state, in childhood, or in adolescence, so the harmful gene is not transmitted to any progeny. Or, he may live to a ripe old age, but either is sterile or never marries. All of these result in removal of the gene from the population. This does not mean that eventually all genes of this type will be eliminated. Mutations are constantly adding new copies of the gene. The input through mutation tends to balance the outgo by genetic death which causes establishment of an equilibrium of the gene in the population. Mutation rates of different genes vary, so the **gene pool** of any specific gene will vary accordingly. Those with a more rapid rate of mutation will naturally be more abundant in the gene pool. Also the size of the gene pool varies according to the rate of elimination by genetic death. Some harmful genes may not prevent persons from marrying and having children. Still they may be harmful enough so that fewer who express it will marry, and those that do may have fewer than the average number of children. Such genes will tend to have a higher frequency in the population than those genes with a 100 percent genetic death for those who express them. Dominant mutations, which cause 100 percent genetic deaths will not become established at all because they eliminate themselves as fast as they appear as mutants. Most harmful mutants, however, are recessive and may become established at a rather high frequency before equilibrium is reached.

Alteration Of The Gene Equilibrium.   Changes in the environment can result in an alteration of the gene pool of specific genes. Genes which were beneficial in one environment may become harmful in another, so selection shifts and elimination increases. The Americans of West African ancestry have a gene pool of only about 8 percent of the gene for sickle-cell anemia as compared to a high of 31 percent in areas of West Africa today. For the past several generations in America, malaria has not been a problem, and selection has been all against the gene. Of course, allowance must be made for an inflow of genes from other races, but this is not sufficient enough to account for all the reduction in the gene's frequency.

Since environments rarely remain constant over long periods of time, there tends to be continual changes in the gene pools. Genes which were once harmful may become less harmful with a resulting increase in the gene pool in a population. The gene pool for PKU in America has an equilibrium frequency of about one percent. About one gene in one hundred at one point on a chromosome is this recessive gene, while the other ninety-nine are the normal alleles. For many generations this gene resulted in 100 percent genetic death when it became homozygous. About one child in ten thousand became homozygous and was so abnormal mentally that reproduction was never achieved. Such removal of two of these genes for each genetic death did not reduce the gene pool, however, because the genes lost were replaced by mutations of the dominant normal allele to the recessive gene for PKU. The mutation rate was about one in five thousand so an equilibrium was established.

Then came the discovery of a way to prevent genetic death of the homozygous recessive persons by withholding the amino acid which was causing the brain damage. The environment had changed, and now these people could be normal and reproduce. The outflow of the gene was greatly reduced, but the input by mutation has remained the same. Hence, the frequency of the gene will increase. In time it may double to 2 percent, and then the cases of PKU will be about four times greater than they are now. Even greater numbers would be expected eventually unless something is done to substitute human selection for natural selection.

This story can be repeated over and over again. As ways are found to prevent the harmful effects of mutant genes, the number of these genes increases in the population. A sad ultimate end to such a process would be a population where almost everyone would inherit severe abnormalities and could be kept alive and reproducing only by continuing medical treatment. To prevent this tragedy there must be a restriction of reproduction by those who are saved from genetic death by medical techniques. This introduces a major moral and social problem which must be faced. If therapeutic abortion ever becomes generally acceptable, such persons could bear only normal children if amniocentesis was performed and embryos with the defective genes were aborted very early in their development.

There is another very important way in which the gene equilibrium can be altered. That is the increase of the mutation input. Since the great majority of mutant genes has a harmful effect, this results in an increase of the gene pool for harmful genes. Man today is being exposed to increasing amounts of agents which have been shown to be mutagenic. High-energy radiation is one such agent. X-rays are used extensively for medical diagnosis and treatment. Radioactive isotopes are used in industrial plants and for generating electricity with some exposure to workers, and non-workers through pollution of waterways with wastes from such plants. Then, there is atmospheric pollution from bomb tests, and this would greatly increase should such bombs ever be used for wartime purposes. All of these tend to increase man's gene pool of harmful mutations. At present the average increase is very small, but the possibility of much larger increases must be kept in mind as we use these sources of radiation in the future.

Now that certain drugs and chemicals have been shown to be mutagenic also, we have the problem of carefully evaluating food additives, preservatives, insecticides, etc. Drug abuse may be a factor which will influence future gene pools.

MONKEYS AND PEOPLE

A study of monkeys showed that .44 percent of their offspring show some kind of abnormality. A similar study of human births showed about 5 percent abnormalities. This includes even small defects. Why should there be such a difference? The difference probably reflects the influence of man's humanity. A baby monkey which is seriously defective will usually die. There are no medical techniques in the jungle to save it. Hence, selection is very strong against genes for such defects. Man, however, has ways of saving many defective babies so the rate of elimination of the harmful genes involved has been reduced. It is also true that there is much more close inbreeding in a society of monkeys, so that harmful recessive genes are brought out and eliminated more readily. In time this results in a reduction of the gene pool for harmful recessive genes. Thus, while everyone would urge a continuation and expansion of man's technique to preserve and make life better for those with inherited defects, the practice of these humane efforts has created problems.

GENE POOLS AS AN INDICATION OF POPULATION ORIGINS

A study of the proportion of different kinds of genes in world populations gives us many important clues as to the origin of the people living in a particular area. The blood characteristics are used extensively in such studies. Results of world wide studies of the blood types has led to the theory that all people were originally type O, then there was a mutation to A. This antigen spread to many parts of the world, but failed to reach the people of northern Asia. The American Indians seem to have migrated from northern Asia to America at this time, and many American Indian tribes are almost entirely type O. Then much later in man's history the B antigen appeared as a mutation in Central Asia, and this is where it shows the highest incidence today.

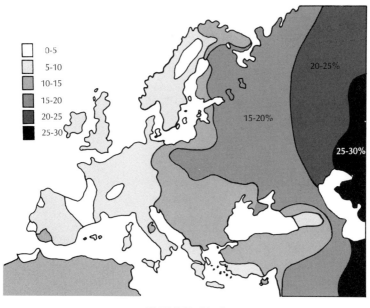

0-5
5-10
10-15
15-20
20-25
25-30

20-25%

15-20%

25-30%

FIGURE 12–1

Distribution of the B blood antigen in Europe and part of Asia. Note how the antigen has spread from an origin in central China.

From here it spread into Europe, and we can actually trace the invasions of Mongols from the east by the residue of genes for the B antigens which they scattered along their invasion routes. Certain isolated regions escaped the invaders. The Basques, on the northern coast of Spain were protected by high mountains, and they are probably more like the ancient Europeans. They have only 1.1 percent B antigens while other regions of Spain have about 9.2 percent. A similar area is found in the Caucasus area of Georgia and Armenia which are parts of the Soviet Union.

Table 12–2 shows the percentage of the blood types and the Rh factor in a number of different world populations. These figures bear out some of the points just made. According to the Rh factor distribution, the gene for Rh-positive must have been originally present in all people, and the gene for Rh-negative appeared somewhere in Europe.

DETERMINING GENE FREQUENCIES IN POPULATIONS

When the frequency of a recessive gene in a population is known, it is easy to calculate the chance of the gene becoming homozygous and being expressed. Let us return to the frequency of the gene for PKU which is established as about one in one hundred.

TABLE 12–2

Variations in Blood Types and Rh Factor in Different Populations

Population Group	Percentages Expressing Blood Antigen Traits					
	O	A	B	AB	Rh⁻	Rh⁺
American Indians (Utes)	97.4%	2.6%	0	0	0	100%
Australian Aborigines	42.6	57.4	0	0	0	100
Basques (Spain)	57.2	41.7	1.1	0	30	70
Hawaiians	36.5	60.8	2.2	.5	—	—
English	47.9	42.4	8.3	1.4	15.3	84.7
French	39.8	42.3	11.8	6.1	17.0	83.0
Germans	36.5	42.5	14.5	6.5	—	—
Italians	45.9	33.4	17.3	3.4	—	—
Japanese	30.1	38.4	21.9	9.7	1.3	98.7
Russians	31.9	34.4	24.9	8.8	—	—
Chinese	34.4	30.8	27.7	7.3	1.5	98.5

Since each child receives two genes of each kind, the chance that the first will be for PKU is 1/100. The chance that the second will be for PKU is the same. Therefore, the chance that both will be for PKU is $1/100 \times 1/100 = 1/10,000$. In most cases we know the number of homozygous persons and want to know the incidence of the gene in the population, so we must work backwards. If statistics show that one child out of ten thousand is born with PKU in the United States, this is the incidence of persons receiving two of the genes, 1/10,000. To find the incidence of single genes for PKU we then need only to take the square root of 1/10,000, and we get 1/100. To find the frequency of carriers of the gene, we must multiply this by two because each person has two genes at this chromosome locus, and there is a 1/100 chance at each locus. Thus, $1/100 \times 2 = 1/50$.

As another illustration, statistics show that about one person in twenty thousand is an albino. Therefore we can take the square root of this figure and find that about one gene in each 141.5 is for this trait. Hence, about one person in seventy-one is a carrier for the gene.

For genes which are more common in a population we usually use the frequency percentage of the gene being considered. For instance, suppose you study representative samples of Seattle's population and find that 9 percent of them have unpigmented (blue) eyes. We can take the square root of this and find that the gene frequency for the recessive gene for blue eyes should be about 30 percent. This means that the rest of the genes at this locus on the chromosomes are for pigmented eyes, various shades of brown. So there must be 70 percent for brown. If we want to know how many of the brown-eyed people are homozygous (carry two genes for brown), we simply multiply the frequency of the gene for brown by itself, $.70 \times .70 = .49$. Thus, we can deduce that 49 percent of those in Seattle have two genes for brown. We already know that 9 percent are homozygous blue, so that leaves 42 percent who are heterozygous brown. This method of determining gene frequencies is known as the **Hardy-Weinberg principle** and has many applications.

These figures assume random, or non-selective mating, and no selective advantage for either color. If, as is probably true to some extent, blue-eyed persons tend to marry each other more frequently than pure chance selection would dictate, then the calculations for carriers of the gene for blue eyes would be somewhat higher than is actually present. Also, if brown-eyed persons have a greater chance for survival than blue-eyed persons, the figures for heterozygous persons would be higher than calculated. And, if the heterozygous persons have some advantage over either of the homozygotes, again the figures would be off. Also, migration can play a part. If more blue-eyed than brown-eyed persons migrate to Seattle from other places, then the estimates will be thrown off. The reverse, of course, is also true. All these factors have to be considered in estimating gene frequencies in populations.

When the heterozygous persons can be recognized, it is possible to check on the results obtained by the Hardy-Weinberg principle to see if some factors besides chance are operating. In Sicily, for instance, it has been found that 20 percent of the population has type N blood (of the MN blood groupings). From here we can calculate the expected percentage of types M and MN. We use the symbols L^N and L^M for the two genes involved.

Frequency of homozygous $L^N L^N = .20$

Frequency of $L^N = \sqrt{.20} = .45$

Frequency of $L^M = 1.00 - .45 = .55$

Frequency of homozygous $L^M L^M = (.55)^2 = .30$

Frequency of heterozygous $L^M L^N = 1.00 - (.20 + .30) = .50$.

We now know the expected frequency of people who are M and MN on Sicily. We can now compare this with the number observed to see if selection, migration, or other factors affect the frequencies:

TABLE 12–3

	M	MN	N
Observed percentage:	32%	48%	20%
Calculated percentage:	30	50	20

The two sets of figures are in such close agreement that we can assume that mating is random with regard to this blood trait and that there is no survival of one over the other, nor are any other factors affecting the frequency.

We find quite a disagreement in the calculated and observed frequencies, however, when we examine a Nigerian population with respect to the gene frequency for sickle-cell anemia. We find that about 4 percent of the babies born have sickle-cell anemia. From this we can calculate the frequency of this gene at 20 percent, which means that the normal allele would be expected to be 80 percent. Homozygous normals would be set at 64 percent, and heterozygous carriers at 32 percent.

It is possible to recognize the heterozygous persons by exposing a drop of the blood to low oxygen. Some of the cells will then sickle. When such tests are made of the normal people, it was found that 46 percent of the total population were *S/s*. This incidence which is significantly higher than the calculated 32 percent reflects the selective advantage enjoyed by the heterozygotes in resistance to malaria. No doubt, many of the homozygous *S/S* persons had died of malaria.

Thus, we see that population genetics can give us much information about human genes and their distribution in different regions of the earth.

GLOSSARY

allele (short for allelemorph). One of a pair, or series, of alternative genes which can lie at a particular locus on a chromosome. The gene for albinism, *a*, is an allele of the gene for normal pigmentation, *A*.

amniocentesis. Removal of some of the amniotic fluid from around the embryo by means of a slender needle inserted through the abdomen of the mother. Used to study embryonic cells for abnormalities.

amnion. A membrane surrounding the embryo; contains amniotic fluid in which the embryo floats.

antibody. A plasma protein, gamma globulin, produced in response to contact with a foreign antigen; reacts with and neutralizes such foreign antigens.

antigen. Substance which can stimulate the production of antibodies. Typically antigens are proteins, but other compounds can also have antigenic properties.

autosome. Any chromosome which is not a sex-chromosome (not an X- or Y-chromosome).

Barr body. A body staining heavily with DNA stains; lies against the inside of the nuclear membrane of female cells. It is a tightly coiled X-chromosome. Synonym: sex chromatin body.

centromere. A constriction on a chromosome which shows most clearly after chromosome duplication since it remains single and holds the chromatics together. Synonym: kinetochore.

chromatid. One of two identical halves of a chromosome after duplication. Show clearly in the prophase of mitosis, because the two chromatids of a chromosome are held together by a single centromere.

chromosome. Thread-like, or rod-like, bodies within the nucleus; contain the genes along with some proteins. Are long and slender during interphase, but through a process of coiling they become shorter and thicker during mitosis.

171

co-dominant. Two allelic genes are co-dominant when both are fully expressed in the heterozygote. The alleles for the A and B blood antigens are examples.

concordance. The degree of similarity between twins, and siblings, according to specific traits. Usually expressed as a percentage.

deletion. The loss of a portion of a chromosome.

dichorionic. When twin embryos have separate chorions.

diploid. Having two of each kind of chromosome; normal somatic cells are diploid since each chromosome in a cell has a mate of the same kind. (Exception: X and Y in males are a pair of different size.)

dizygotic. Refers to twins who originate from two different zygotes (fertilized eggs).

DNA. Deoxyribonucleic acid, the material of which genes are made.

dominant. Refers to one of an allelic pair of genes which is expressed largely to the exclusion of the recessive allele of the pair. A trait which appears in a person who is either homozygous or heterozygous for the gene.

drift, genetic. Change in the frequency of genes in a population as a result of chance fluctuations. Usually results when a small isolate does not have genes representative of the population from which it came.

epistasis. When a gene at one locus suppresses the action of another gene, or genes, at different loci. The gene for albinism is epistatic to the non-allelic genes which govern the intensity of melanin deposits.

expressivity, variable. When the expression of a trait resulting from a specific gene is variable. In some cases variations in the environment are responsible; in other cases genes at other loci may cause the variation.

fetus. The human embryo during the final seven months of its prenatal existence.

gamete. A male or female reproductive cell, a sperm or an egg.

gene. The basic unit of heredity; occupies a fixed locus on a chromosome; a unit of DNA which codes the production of a polypeptide chain.

genetic death. When a person fails to transmit genes to future generations. Can result from death before maturity, or failure to reproduce after maturity.

gene pool. The total of all genes in a population.

genotype. The type of genes in an individual. (Compare with phenotype.)

germplasm. The reproductive cells and their immediate precursors.

globulins. Proteins in the plasma of the blood. Alpha, beta, and gamma globulins have been identified.

haploid. A term referring to a single complete set of chromosomes. Sperms and eggs are said to be haploid because they have only one chromosome of each kind.

haptoglobin. An alpha globulin in the blood plasma which binds and carries free hemoglobin from broken down red blood cells.

hemoglobin. Iron-containing protein within red blood cells that transports oxygen. Central core of iron compounds with two pairs of alpha and beta polypeptide chains.

hermaphrodite. One possessing both male and female reproductive organs; has somewhat intermediate secondary sex characteristics.

heterozygous. Differing with respect to a particular pair of allelic genes. Example: A person with the genes A/a is heterozygous for the recessive gene for albinism. Noun: Heterozygote, one who is heterozygous for a particular gene.

homozygous. Alike with respect to a particular gene. One expressing the recessive trait, albinism, is homozygous a/a. Noun: Homozygote, one who is homozygous for a particular gene.

implantation. When an embryo becomes attached to, and embedded within, the uterus.

intermediate inheritance. When both genes of an allelic pair are expressed to an approximately equal degree, resulting in a trait about half way in between the two homozygotes. Synonym: lack of dominance.

intersex. One showing characteristics of both sexes with respect to the secondary sex characteristics.

karyotype. Chart made from a photograph of the chromosomes in a single cell. The chromosomes are cut out and arranged in matched pairs according to length.

kinetochore. Synonym of centromere.

lethal gene. A gene which, when expressed, results in death either in the embryonic state, or shortly after birth.

linked genes. Genes which are located on the same chromosome, yet are not alleles.

locus, pl. loci. A position on a chromosome occupied by a certain gene or one of its alleles.

meiosis. A series of two cell divisions accompanied by only one gene and chromosome duplication; reduces the chromosome number from the diploid to the haploid.

mitosis. Cell duplication; chromosomes and genes are first duplicated, then the double chromosomes separate and are segregated into nuclei at opposite ends of the cell. Cleavage of the cell results in two daughter cells.

monochorionic. When twin embryos share the same chorion.

monsomy. When one chromosome of a diploid set is missing. A monosomic human somatic cell would have only 45 chromosomes; all are in pairs but one.

monozygotic. Refers to twins who originate from a single zygote, fertilized egg.

mutation. A change in a gene which causes the gene to have a different effect. (Some use the word to refer to changes in chromosomes as well as the genes; in this book chromosome aberration is used.)

non-disjunction. When two paired chromosomes fail to disjoin during meiosis. Both members of the pair go to one cell and none to the other.

oocyte. Cell that undergoes meiosis to form an ootid, which becomes an egg, and three polar bodies. Primary oocyte is the first cell, and it forms a secondary oocyte after the first division of meiosis.

oogenesis. The process of egg formation through meiosis of a primary oocyte.

operator gene. A gene which controls the functioning of a group of related structural genes.

overdominance. When a dominant gene has a normal or beneficial effect when it is heterozygous, but can have extremely harmful effects when homozygous.

pedigree. A chart showing the ancestral history of a person for a particular trait, or for a number of traits.

penetrance. The proportion of persons who express genes which they carry. A dominant gene with complete penetrance is expressed by all who carry it. A dominant gene with 80 percent penetrance is not expressed by 20 percent of those who carry it. This is incomplete penetrance. Also, applies to the expression of homozygous recessive genes.

phenocopy. An environmental duplication of a trait which may be caused by heredity in other instances.

phenotype. Traits expressed by an individual. Blue eyes, colorblindness, type A blood, and albinism are examples of phenotypes.

pleitrophy. When a single gene influences a number of different traits.

polar body. Small body given off from the primary and secondary oocytes in the unequal division of these cells during meiosis.

polypeptide chain. Chain of amino acids assembled at the ribosomes as coded by messenger-RNA from the genes. May be a protein or combine with other chains to form a protein.

recessive. Refers to a gene which is not expressed when heterozygous with a dominant allele. Traits are called recessive when due to recessive genes.

regulator gene. One which produces a repressor which prevents the operator gene from stimulating functional genes into action.

ribosomes. Small spherical bodies in the cytoplasm; the site of protein synthesis; made of protein and RNA.

RNA. Ribonucleic acid, a nucleic acid which carries codes for specific tasks. Messenger-RNA carries codes from genes to ribosomes; transfer-RNA codes specific amino acids and moves them to the ribosomes; ribosomal-RNA attracts the messenger-RNA. Other types are also known.

sex chromatic body. Snyonym for Barr body.

sex chromosome. One involved with sex determination; the X-chromosome and the Y-chromosome are sex chromosomes.

sex-linked. Refers to genes which lie on the major portion of the X-chromosome, the portion not homologous with the Y-chromosome. Traits resulting from sex-linked genes.

sibs, siblings. Brothers and sisters of the same parents; born at different times; not twins or other multiple births.

somatoplasm. All of the body except the reproductive cells and their immediate precursors. Opposite of germplasm. Adj.: somatic.

spermatid. Cell formed by spermatogenesis; becomes a sperm.

spermatocyte. Cell that undergoes meiosis to form sperms. Primary spermatocyte is the first cell, and this forms two secondary spermatocytes in the first division of meiosis.

teratogenic. An agent which results in an abnormal embryo when it affects the pregnant mother.

transferrin. Beta globulin of blood plasma which transports iron.

translocation. When a portion of a chromosome, or an entire chromosome is attached to another chromosome.

trisomic. A condition where one chromosome is present in triplicate rather than the normal pair in somatic cells. Trisomy-21 means that there are three chromosome-21's rather than the normal two.

X-chromosome. One of the sex chromosomes; associated with sex determination. Normal females have two X's, normal males have one X.

Y-chromosome. One of the sex chromosomes associated with triggering male characteristics. All males have a Y-chromosome.

zygote. A cell produced by the union of a sperm and an egg. A diploid cell with the potentialities of developing into a human being.

INDEX